Managing Electronic Media

ELECTRONIC MEDIA GUIDES

Elizabeth Czech-Beckerman *Managing Electronic Media*

Ken Dancyger *Broadcast Writing: Dramas, Comedies, and Documentaries*

Donna L. Halper *Full-Service Radio: Programming for the Community*

Donna L. Halper *Radio Music Directing*

William Hawes *Television Performing: News and Information*

Robert L. Hilliard *The Federal Communications Commission: A Primer*

John R. Hitchcock *Sportscasting*

Andrew F. Inglis *Satellite Technology : An Introduction*

Carla B. Johnston *Election Coverage: Blueprint for Broadcasters*

Marilyn J. Matelski *TV News Ethics*

Marilyn J. Matelski *Daytime Television Programming*

Robert B. Musburger *Electronic News Gathering: A Guide to ENG*

Managing Electronic Media

Elizabeth Shimer Czech-Beckerman

Focal Press
Boston London

Focal Press is an imprint of Butterworth–Heinemann.

Recognizing the importance of preserving what has been written, it is the policy of Butterworth–Heinemann to have the books it publishes printed on acid-free paper, and we exert our best efforts to that end.

Library of Congress Cataloging-in-Publication Data
Czech-Beckerman, Elizabeth Shimer.
Managing electronic media / Elizabeth Shimer Czech-Beckerman.
p. cm.—(Electronic media guide)
Includes bibliographical references.
ISBN 0-240-80072-9 (paperback)
1. Broadcasting—Management. I. Title. II. Series.
HE8689.6.C94 1991
384.54'068—dc20 90-40738
CIP

British Library Cataloguing in Publication Data
Czech-Beckerman, Elizabeth Shimer
Managing electronic media.
1. Electronic equipment. Applications—(Electronic media guide)
I. Title. II. Series
621.381
ISBN 0-240-80072-9

Butterworth–Heinemann
80 Montvale Avenue
Stoneham, MA 02180

10 9 8 7 6 5 4 3 2 1

Printed in the United States of America

Contents

Preface

This book presents an overview of management's responsibilities as related to the organization, principles, and practices common to radio, television, cable, and related electronic media. Although geared toward broadcast and media management from a beginner's viewpoint, the book has also been designed to be of use to individuals already holding entry-level positions and planning to move to upper management, or possibly, ownership.

The scope of this book is limited to aspects of the American systems. However, the serious, aspiring professional should also examine other national media, not only because of technical and philosophical differences, but also because many American media conglomerates have important financial interests elsewhere both in electronic media and in print.

Material for this book is based upon my 30 years of practical and management experience in commercial and noncommercial broadcasting as well as my ongoing professional research and on the literature necessary for my teaching at the college level since 1959. In addition, through continuous correspondence with many former students who have moved into media decisionmaking positions, I have been kept current with the commercial field and sensitive to what young people at the entry levels want and need to know.

The history of mass media reveals there has always been a close relationship among business, technology, and culture. As illustrated in Chapter 7, the model of this book has been formed by those linkages. There are a number of ways to look at management. One should be familiar with a variety of management styles, especially which style fits best within a particular corporation in what is referred to in management literature as *corporate culture*. Just as no one family is exactly like any other family, no one corporation and its personnel work together exactly like any other corporation. For this reason, it takes time for a newcomer to become oriented to written policies and procedures as well as to the "unwritten" rules that often must be observed and sensed, since they are usually unspoken.

Although the road to leadership begins with education, it may take 25 to 35 years before one can sit in the general manager's chair. To rise to the challenge of corporate leadership requires skills with people as well as extensive knowledge about the external environment in which a business exists, such as competition, possible takeovers, and technological changes. For this reason, the focus of this book is to introduce background and information that management would like its employees to possess.

Whereas entry-level positions are most likely to be skill- and task-oriented, managerial positions require a thorough understanding of the industry, as a whole, as well as the philosophy of ownership of the particular station or system. Adaptability to rapid change is the hallmark of a successful media manger, especially in today's period of technological innovations. Besides radio and television, which distribute audio and video broadcast into the ether, there are a number of other electronic systems that may interact, boost, redirect, or reshape the original broadcast signal. These related technologies, which include cable, satellites, computers, fiber optics, and digital processes, are discussed throughout the text from management's viewpoint. Because job advancement opportunities often appear in such delivery systems, this book also includes some background and explanation of possible alternate career moves for the ambitious person. Through conglomerate ownership, many such systems also offer management training and job opportunities.

As summarized in Chapter 1, from the very beginning broadcasting has been a commercial enterprise, with stations "hooked together" by the telephone company's long distance lines to create networks, which eventually were supported primarily by advertising as their main source of revenue. Noncommercial broadcasting, originally designed for instructional and educational purposes, began in the same decade. Arrival of television and cable, while different technologically, basically maintained the same organizational structure of radio, as do many of the new technologies and satellite systems in use today.

An important chapter, written for the person with little or no technical background, Chapter 2 explains spectrum as related to AM, FM, television, cable, and satellites; and shows how such basic knowledge is essential not only to engineers and to management for regulatory purposes, but also to sales, programming, production and financial matters.

In Chapter 4 the relationship between media economics and organizational structure is examined, with particular emphasis upon advertising, advertising agencies, rates, how time is bought and sold, and markets. Without advertising, the majority of programs could not be produced or aired. Even on noncommercial stations there is a modified form of "soft" advertising by giving air credit to donors and supporters. This chapter is of special value to people interested in broadcast sales, which is normally the major avenue to top management positions.

Chapter 5 illustrates program practices and challenges that programmers face every day, particularly strategies of program placement as to daypart and time of airing. These terms and practices are also used by cable systems, syndicators, and program producers when buying time or setting rates. In addition to identifying basic program classifications, this chapter examines the role of program series and of syndication as seen from the program director's viewpoint in addition to types of audience research available to assist in decisionmaking. This chapter also explores the dilemma of trying to serve the public and, simultaneously, to earn enough money to produce programs of quality that are likely to succeed. With so many new delivery systems today there is a shortage of both innovative programming and program approaches. It is said that the 1980s was the decade of new technology, and that the 1990s need to catch up with new programming techniques that are also economically viable.

The broadcaster operates with many forms of control to consider. Chapter 6 reviews the many types of control, including regulatory and nonregulatory. The 1980s was a decade of deregulation in many ways, but regulation still exists in matters of technical quality, licensing, antitrust laws, matters of unethical practices, and First

Amendment rights. Other external controls may be exerted by public tastes in the form of pressure groups, citizen groups, or boycotts of advertisers whose commercials appear on programs that viewers consider to be offensive. Efforts at self-control by the media are seen through the formation of professional and industry codes or guidelines. There is also possible a form of source control if reporters fail to investigate stories supplied by the source itself and which, therefore, may be biased. Management must have its finger on the pulse of all such matters because failure to do so may adversely affect station economics or involve it in legal complications with various governmental or public bodies.

Fundamental management principles and theories, discussed in Chapter 7, offer the beginner a survey of managerial styles and purposes. This information enables employees to understand and assess the managerial style of the person under whom they presently work, and also challenges a neophyte to consider how she or he would do it differently if the roles were reversed. The serious future manager will study these and many other theories in depth as part of self-preparation for the future. Broadcast management requires a broad knowledge of business administration, personnel, law, psychology, sales, programming, technology, and people, with enough flexibility to modify managerial style, adapting to changes in the constantly shifting media industries anticipated in the 1990s.

Chapter 8 was written especially for the individual seeking entry-level employment in the industry, and offers guidelines for career preparation: how to find a job, prepare a résumé, and how to handle auditions and job interviews. Brief descriptions of activities and responsibilities found in different departments help a newcomer to appreciate the teamwork necessary between departments as well as to find an initial area of interest for the first position. Cable systems have similar organizational structures, though they are not as elaborate and use different job and department titles. This chapter also has a section concerning advancement prospects, and suggests the numerous media-related job opportunities that exist in organizations and systems outside of broadcasting.

It is important to keep in mind that corporate culture is continually modified by changing internal and external cross-cultural interactions and influences. Suggested additional readings are found in the bibliography at the end of this book.

Not only am I grateful to mentors of the past and the many professional colleagues who have shared their knowledge and support, but I especially thank the hundreds of students who have kept in touch after they graduated, sending me current material related to their jobs and to the field, making comments as to which teaching material was particularly helpful to them, or offering suggestions about new units to add to courses. I have discovered that it is very true: "By your students you'll be taught." Finally, I wish to thank my dear husband, George, who patiently supported me in my endeavors to write this book.

Introduction to Broadcasting and Electronic Media

BROADCASTING AS A BUSINESS

We cannot understand the present until we know the past. And the beginning is, in a sense, the story of the three "Rocks": NBC (30 Rock), ABC (Hard Rock), and CBS (Black Rock).[1] The first broadcast receivable by the public was on KDKA, on November 2, 1920, when Frank Conrad, an engineer for Westinghouse in Pittsburgh, made test transmissions at an amateur radiotelephone station by playing phonograph records. Other amateurs who could pick up his signal began to request music, and then sports and news.

Seeing an opportunity to make money by selling the new "amateur wireless sets," Horne's department store in Pittsburgh stocked ready-built receiving sets and ran an ad in the paper stating, "Air Concert 'Picked Up' by Radio Here." And so business and broadcasting were united.

A group of manufacturers of radio receivers (General Electric, RCA, and Westinghouse), which called itself the "Radio Group," saw this new technology as a way to air programs to sell their radio receivers. The Radio Group owned most equipment patents, with WJZ as its flagship station.

Meanwhile, competition arose from the "Telephone Group," which consisted of AT&T, Western Electric, and station WEAF. Because this group controlled the telephone lines, its philosophy was to initiate what it called "toll broadcasting" of sponsored programs.

The Radio Group wanted to air public messages on many stations in order to stimulate the market for consumer goods, whereas the Telephone Group had a sender's emphasis and wanted to own only a few stations, with no responsibility for content and full control of all advertising or distribution of personal messages. The battle lines were drawn.

In those early years radio stations began to hook up informally into networks by renting telephone lines for this purpose. Seeing that the public preferred the Radio Group's philosophy, cross-licensing agreements were made in 1926, according to which AT&T of the Telephone Group would continue to connect all stations for broadcast purposes, and RCA would develop commercial network broadcasting.

Originally the Radio Group had hoped that radio would offer programs with no advertising. But when the Radio Group realized that selling radio receivers no longer brought in enough revenue to support the radio stations, radio became an advertising medium in 1928.

Birth of the Networks

At that time National Broadcasting Company(NBC) had two networks, designated the red and the blue for the red and blue lines drawn on engineering maps to indicate which phone lines hooked up which stations. Throughout the depression, radio was the major source of entertainment for people who could not afford to go to movies or theaters. Columbia Broadcasting System (CBS) entered the picture through its talent-booking strength, and made stars of many of its early performers. Mutual Broadcasting System became the third network in 1934, with WOR as its flagship station. MBS chose to remain a radio network and continues so today. In 1942 the FCC's chain broadcasting rules, which prohibited an entity from owning more than one network, forced NBC to sell one of its two networks. NBC kept the red and sold the blue to what became the American Broadcasting Company (ABC).

Business was integrated with broadcasting from the start. RCA, which owned NBC after GE and Westinghouse withdrew ownership, also had control of the Victor Talking Machine Company, and expanded to manufacture radio equipment, phonograph records, recording systems, projectors, sound equipment, public address systems, and concert-booking services.

CBS began in 1927, eventually was named the Columbia Phonograph Broadcasting System, and retained the name when the record company withdrew from ownership. It was William S. Paley who purchased what he called a "patchwork, money -losing little company" and turned it into a strong success by originating the network-affiliation concept and innovative programming.

In the 1930s programming became highly developed. Through the public's interest in the news of World War II happenings, newscasting became an important program commodity. Comedy programs and soap operas dominated the evening schedules, in addition to live musical programming, both popular and classical. The depression enabled networks and national advertisers to dominate program control because at that period they were the only groups with big money.

When television became a force in the 1940s, it began to "steal" big radio talent with lucrative contracts, borrowed radio formats and changed them into visual forms; and the networks switched more financial support to television than to radio. The major radio networks eventually became also the major television networks and remain dominant today, even though we have additional networks.

Network Stations

Owned-and-Operated Stations It was development of the network system of station ownership and affiliation that led to the establishment of broadcasting as a national industry. Even today it is the primary function of radio and television networks to distribute programs to local stations around the country. Those stations may be either owned-and-operated (O & O) or affiliated. The networks not only own the facilities of their O & Os but also provide the programming, operate the station, and hire and pay the management and staff.

According to current FCC rules and regulations, the maximum number of stations that may be owned by an individual, group, network, or conglomerate are 12 AM, 12 FM, and 12 TV stations, as long as the total audience reached does not exceed 25% of the nation's homes. The purpose of such limits is to prevent excessive control of content and to spur competition in order to provide diversity of programming. Exceptions to the maximum are discussed in Chapter 3.

O & O stations establish a network's prominence in the major markets, generate substantial local income for the network, train potential managerial candidates for future network positions, and provide the flexibility which allows a network to experiment with different program ideas or to test new equipment in one or more specific markets.

Affiliates An affiliate is a station that is not owned by a network, but desires to be associated with a network for financial and programming advantages. Network affiliates cover markets that the O & Os may not be able to cover. Through a combination of O & Os and affiliates, the networks try to have their programs aired in all of the 212 U.S. markets, which is discussed in more detail in Chapter 4. Understandably, the networks seek to own or affiliate with the most successful stations. Affiliates also share some of the network's earnings from national advertising.

Independent Stations Not all stations wish to affiliate with a network. Some may prefer different target audiences or to air programs with more local appeal, depending upon local advertising for their financial support. Although independent stations face obvious challenges when competing with networks, O & Os, and affiliates, they also have more flexibility to experiment with creative programming as well as to change corporate decisions more rapidly. Quite often the independents seek to serve a particular ethnic audience, which may have selective cultural, entertainment, or economic interests different from mass markets.

Television

Growth of the Television Networks By 1960 more than 500 television stations were on the air, of which about 90% were affiliated with one of the three networks. Full color production on networks began in 1966, but not until about half of the U.S. television homes had color sets did color become the norm in 1972.

Introduction of videotape recorders by the end of the 1950s initiated the demise of much live television programming as television turned to film and videotape for production and distribution.

Until the 1960s UHF stations could not be received on home television receivers. But audiences for UHF grew with the FCC's passage of the All Channel Receiver Bill, which required that all television sets manufactured after 1964 be able to receive UHF as well as VHF stations. This strengthened ABC, whose O & Os were UHFs, and finally made all three networks equally competitive. The new technology of lightweight, portable cameras and of microwave transmission introduced electronic television news gathering by the mid-70s. Coupled with satellite capability to hook up news sources, television gained the ability to go anywhere in the world and to provide instant coverage of important news, sports, and cultural events. This ended the era of prerecorded or filmed inserts in news programs, and changed news from a loss-leader to an important source of network income. Advertisers flocked to buy time in news programs.

However, the introduction of new technologies, as discussed in detail in Chapter 2, began to have a negative impact on the networks in the 1980s; and the networks' dominance began to fade. By 1990, networks' share of the audience had dropped from an average of 90% to just 67% with the new technologies taking the remaining shares. Some of the non-network viewers were watching independent, nonaffiliated local stations; some turning to public television. Additional technologies causing audiences to turn away from regular television were cable, pay-TV on cable, prerecorded video cassettes, and video games.

Networks in the 1990s Costs of operating a network and developing programs rose astronomically by 1990 and flattened out the profit curve. For example, CBS reported sales of $2.5 billion, but made a profit for investors of only $60 million.[2] To overcome their fiscal problems, the networks devised new strategies to shore up their profits and their programs by joining with the forces that were nibbling away at their audiences. ABC and CBS set up divisions to create special programming for satellite distribution to cable systems across America. In addition, networks were in the process of asking the FCC to permit networks to own cable systems; the inquiry was unsettled at this writing.

Until the decade of the 1990s, a national network service meant the directionless broadcasting, usually by satellite, from a network source to a local television station that rebroadcast the program to its local audience. Today the term *network* can also mean transmission of programs from the network source to a satellite for retransmission to cable systems, with a number of channels wired directly to subscribers' homes by means of cable, with the capability of providing a greater number of program choices, compared to the few channels of the networks. In addition, despite smaller audiences, cable's lower costs of distribution create the potential for significantly higher profits.

It will be interesting to see how the networks face the new challenges. Will their programming become more original? Will they be able to continue to attract advertisers willing to pay for expensive time? Will they find a way to integrate with the technologies? Stay tuned to the 21st century for the answers to network management's dilemma.

Independent Television A fourth network, Fox Broadcasting, born in the 1980s, introduced strength for the independents and a challenge to the established networks. By developing and producing its own special programs and miniseries and distributing them via satellite to *ad hoc* networks of selected independent stations, Fox not only created the flourishing syndication marketplace but also, along with cable, began to make inroads on the three major networks.

Radio's Response to Television

Programming After television achieved national coverage and borrowed many of radio's programs, including soap operas, radio adapted during the mid-40s through the mid-60s by changing formats to *hit radio*. As television took over radio's former mass programming, AM radio became more a local medium that featured personalities and music.

FM was initially slow to receive public acceptance because people who wished to hear FM had to buy expensive separate FM tuners to pick up FM signals. During this transition, AM owners often obtained FM licenses and then either duplicated the AM programming on FM or used FM to program classical music, because FM offered much better audio quality than did AM. Eventually radio receivers were manufactured to pick up both AM and FM, at which time the FM programmers broke into other formats, which included rock and popular sounds as well as classical. FM gradually replaced AM as the preferred music medium, leaving AM to turn to news, talk, information and non-music formats.

When ABC decided to split its own radio network into four different sounds in 1968, the network introduced another new radio network concept. Today, ABC radio offers seven different formatted networks: Contemporary, Direction, Entertainment, FM, Information, Rock, and Talk Radio. This was the beginning of the move by radio toward focusing on local target audiences. As a result, only about one third of the nation's radio stations kept their network affiliations, with two-thirds offering local programs instead. However, in the 1980s satellite connections and ancillary technologies, plus interested advertisers who found that today's network radio is a good and fairly reasonable way to reach selected target audiences in the entire country, brought about a revival and growth of the number of radio networks.

Specialized Radio Networks The diversity of radio network offerings continued to expand in the 1990s; and at this writing there are more than two dozen radio networks that feed not only 3-minute programs, newscasts, and features but also music specials, sporting events, news/talk programs, and drama.

The number of radio stations continues to increase. 1990 saw the existence of 4932 commercial AM stations, 4155 commercial FMs, and 1374 noncommercial FMs. If the FCC allows expansion of the AM band to 1705 KHz at the upper limit, there will be room to accommodate even more AMs.[3]

Use of Satellites and New Technologies Thanks to satellite transmission today "instant" radio networks are sometimes formed by program syndicators in conjunction with advertising agencies. Such networks transmit programs only occasionally, and their affiliates can vary from week to week. Sometimes, radio sales firms will put together such an "unwired" national network for those advertisers willing to sponsor different shows in dozens of cities. In additional, network executives have learned from the success of radio syndicators that concert specials invariably improve the ratings.

Network radio is not necessarily low budget. For example, CBS radio paid $12 million for the 4-year national radio rights to the NFL games. In general, however, most agency executives think network radio's future is assured because of (1) its comparatively low cost compared to television, (2) its frequency values, and (3) its ability to attract and keep large numbers of listeners whose age, lifestyle, gender, income level, purchasing habits, and ethnic origin match the specific demographics required by certain advertisers. Big advertisers find that radio networks help to boost their corporate image. Some advertisers sponsor entire programs; others run advertising spot campaigns.

Today there are approximately ten radio stations for one television station. But in the radio network picture, what makes or breaks a radio network is the quality of the affiliates, not just the sheer number of them.

With the new technologies, radio networks appeal to advertisers in the same way that magazines and cable television do, that is, by programming to particular demographic groups which are difficult to reach through network television.

NONCOMMERCIAL BROADCASTING

Purpose

Only the name has changed! What is now called Public Broadcasting was originally established as Educational Broadcasting. In 1925 educational institutions and groups held 171 AM licenses. By 1934 there were still about 25 on the air. Later, in 1941, when FM was authorized by the FCC, five FM channels were set aside by the FCC especially for educational purposes. Later, in 1945, with the revision of frequency allocations by the FCC, twenty FM channels between 88 and 92 MHz were reserved for noncommercial educational FM stations. With the arrival of television, some 242 channel assignments were reserved for educational purposes, of which 80 were VHF and 162 UHF. Today the latest FCC table contains 615 educational assignments.

Although these stations originally served only educational needs, they gradually added cultural and public interest programs at the local level. When they eventually created their own network, they also developed special noncommercial network programs.

The name changed to Public Broadcasting as a result of passage of the Public Broadcasting Act of 1967, which authorized establishment of the Corporation for Public Broadcasting, organized primarily as a way to provide federal funds to qualified stations. In 1971 National Public Radio was formed to provide interconnection and programming services for public radio stations. Meanwhile, public television stations, under the name of Public Broadcasting Service (PBS), offered interconnection and program distribution services starting in 1970.

Whereas commercial broadcasting is supported by advertising, public broadcasting is funded by a combination of federal grants and subsidies from corporations and loyal audiences.

Delivery Systems

It is noteworthy that PBS pioneered the use of satellites to replace A T & T telephone lines, coaxial cables, and microwave to relay network signals to affiliates when public television stations began to receive their network feeds via Westar I. Joining its television sister, National Public Radio (NPR) also started to use satellite delivery for its radio interconnection in 1980, and originated programming four audio channels, initiating the concept later to be followed by commercial network's use of satellites.

Some noncommercial networks hook up for special informational or entertainment programs, the best known of which is National Public Radio's "All Things Considered," the 90-minute early evening news show that won both the Dupont and Peabody awards, in addition to having built the cornerstone around which NPR has developed its network, which also offers quality classical and jazz programs.

CABLE

Cable television was the first competitive technology to challenge television broadcasting. In fact, by 1990 approximately 58% of the nation's homes were cable subscribers, and cable was an $18-billion-dollar industry. This reflects a quantum jump form the 1950s when Community Antenna Television (CATV) systems began to provide television signals to residents of small, isolated hilly communities unable to receive television signals directly.

Impetus to CATV's development resulted indirectly from the FCC's 1948–1952 freeze on new television station construction. Just as in the early days of radio, early cable systems were mom-and-pop type, with only five or six channels, reaching no more than about 1000 homes with a simple system which consisted of a large antenna on top of a hill to pick up the closest television stations, and then amplifying and feeding those signals to home receivers via cable wired into homes.

Unlike broadcasting, cable earns its revenue from subscriber fees rather than from advertising. To gain more subscribers and greater revenue, cable began seeking subscribers in large cities by offering improved quality signal reception, providing addition signals, originating programs on unused channels, and by initiating pay systems for special programming.

In the beginning, television broadcasters welcomed cable because it distributed their signals (and advertising) to additional audiences, and also improved their signal quality in poor reception areas. However, by the mid-60s the new cable "enabler" began to appear in a different light when it retransmitted distant television stations' signals, sometimes deleting the signal of the local station. In this circumstance a local broadcast affiliate's programs could be duplicated by cable, which also picked up the program from a distant station, thus resulting in smaller audiences for the local affiliate and its advertisers.

This practice of importing distant signals obscured the natural fixed market boundaries set by the limitations of over-the-air broadcast signals, which not only fragmented the broadcast audiences but also lowered station ratings, which decreased rates that local stations could charge. Thus broadcasters began to see a loss of audience and of revenue. We must remember that cable did not pay the stations for the programs they aired, yet, in a sense, sold the programs to cable subscribers. This practice led to broadcasters' seeking some type of regulation of cable, as discussed further in Chapters 3 and 6.

Noncommercial educational television was also adversely affected by signal importation when cable brought in noncommercial programming from other sources. UHF stations also lost audiences through importation. Keep in mind that the signals of UHF stations generally do not travel as far as VHF, and at that time, the FCC was trying to encourage UHF development. Audience fragmentation was especially hard on UHF's already limited audiences. To eliminate these problems, protective regulations went into effect in 1972. That was reversed, however, in 1980 when the FCC dropped the cable programming rules which had banned importation of most distant signals and had protected some broadcast syndicated programming from cable competition. By that time, through satellite-distribute services, cable systems had become outlets for a wide choice of national program services and, therefore, provided the program diversity that the FCC had always encouraged. As a result, the FCC discontinued protecting broadcasting from cable and encouraged competition between the two.

The 1980s found cable offering a variety of special services in addition to carrying radio and television signals. Cable began to offer programs beyond basic levels to include pay channels, pay-per-view, and interactive two-way capability for such activities as home shopping, opinion surveys, and even home surveillance systems.

CABLE NETWORKS

Besides major broadcasting networks, many states and regions also have their own radio or television networks, some for special purposes such as education, farm information, etc. In addition there also are *ad hoc* networks hooked up for one-time-only special events or entertainment.

When cable systems expanded beyond the mom-and-pop stage and began to enter large markets, big corporations saw the economic benefits of cable ownership and began to buy out existing small systems and to bid for new franchises in large markets. In a sense this followed broadcasting's beginnings and formed *cable networks* owned by multiple system operators (MSOs).

Superstation networks came into being as a result of the superstation concept pioneered by Ted Turner when he innovatively used satellite and cable connections to obtain a national audience for his small Atlanta UHF station.

The 1990s, therefore, found several forms of cable network systems: basic, pay cable, and superstations. Now there are more than 100 cable networks that serve special interest audiences and include such services as Cable News Network (CNN), Black Entertainment Network (BET), Spanish Information Network (SIN), Entertainment and Sports Programming Network (ESPN), Music Television (MTV)—all of which are found in basic cable services.

Among noteworthy large pay cable networks are The Disney Channel, Home Box Office (HBO), The Playboy Channel, and BRAVO (which offers such performing arts as films, operas, symphony performances, and special concerts). For current listings of the many broadcasting and cable networks see the professional trade literature found in the bibliography of this book.

SUMMARY

By reviewing the history of the development of broadcasting, it is clear that from the very beginning broadcasting was a commercial enterprise. The interrelationships between broadcasting and telephone lines helped shape the industry in its present form. The same questions that arose in the 1920s are similar to those being asked today. And the ongoing questioning about the purpose of free broadcasting continues in lively debate.

It was the union between equipment manufacturers and the necessary "long lines" of the telephone company that gave birth to the network system as a mass medium.

In matters of program development, the early years of broadcasting reveal that even then social and economic forces determined the type of programs the public wanted; and the value of free broadcasting to a country suffering from an economic depression is obvious. The philosophy born from those years stirs again in the rising competition between pay cable and "wireless" broadcasting.

While it was development of the networks that contributed to present station organization forms, it was local radio that supported a strong sense of localism in communities. The roles of network O & Os and affiliates, as contrasted to independent stations, continue in both radio and television today.

The impact of the new technologies reveals the management and economic challenges to media managers of the future, and introduces concepts that television net works, in particular, have devised to survive.

When radio's mass audience was siphoned off by that "new" medium, television, radio countered with new formats and found different audiences, even using satellites for interconnection to form "instant" networks, appealing to advertisers interested in reaching new target audiences. The support of advertising continues to be apparent.

Noncommercial broadcasting has expanded from its original intention to "educate." Although still offering instructional material during much of the daytime hours, public broadcasting has developed a strong following of people whose primary interest revolves around in-depth presentation of programs that would not attract large mass audiences, but do attract sizable audiences seeking alternative programming offering enriched coverage of thought-provoking talk and news shows as well as dramas, classical and jazz music, and popular news magazines.

Sometimes new ideas may be generated by looking at the old. This chapter provides a way for management to look at the future through the eyes of the past.

Cable television earns its revenue from subscriber fees, whereas broadcasting's earnings come from advertisers who buy broadcast time to air their commercials.

Broadcasters welcomed cable in the 50s and 60s because cable extended and sometimes improved broadcast signals and enlarged broadcasting's audiences. However, by the 1980s cable became a strong competitor of free television and radio due to the importation of distant signals, the creation of cable networks via satellites, and the provision of a variety of services and programs not available from free television. While such program diversity was welcomed by the FCC, it resulted in reduced viewership and ratings for broadcasting. As a result the FCC has considered offering broadcasting some type of protective regulation in the 1990s, to be discussed later in this book.

Notes

1. Nicknames of the three net works: NBC at 30 Rockefeller Plaza is "30 Rock"; CBS's black granite facing inspired the name "Black Rock"; and ABC is remembered as "Hard Rock" thanks to its popularization of rock music.

2. A talk by Jay Kriegel, senior vice president of CBS, November 16, 1989 during his Earl Wynn lecture at the Department of Radio, Television and Motion Pictures, University of North Carolina, Chapel Hill.

3. *Broadcasting/Cable Yearbook*: 1989.

2

A Spectrum Primer and Related Technologies

Although it is not necessary that owners of broadcast facilities and other electronic media or their managerial teams be engineers, it is helpful for them to have a broad understanding of the range and limits of spectrum; as well as how it ties in with licensing, economics, sales, programming, production, regulation, and related technologies. Without being too technical, this section provides a "working primer" of spectrum as related to broadcasting, "wireless cable," and satellites.

SPECTRUM ALLOCATION

Bands

According to international treaty, the spectrum is broken into eight *bands* as illustrated in Figure 1.

Different bands are set aside for specific services. As seen in the figure, AM operates on medium frequencies (MF), and television uses very high frequencies (VHF) as well as ultra high frequencies (UHF). We find that frequency modulation (FM) is broadcast on VHF as is the audio for television. Satellites use VHF, UHF, and sometimes super high frequency (SHF).

To assure that the usable limited amount of spectrum is apportioned equitably and, also, to prevent overlapping of broadcast signals, there is international agreement as to who will use which portions of the spectrum, and how they are to be used. This is determined by a group of some 130 countries that belong to the International Telecommunication Union (ITU), which establishes on a worldwide basis which bands of spectrum frequencies will be used for a particular purpose. Then each country decides how it will assign the portions allotted to it. The ITU also assigns the initial call letters to be used for radio stations in each country, and has designated the letters W and K to stations in the United States.

Through the Communications Act of 1934, Congress created the Federal Communications Commission (FCC) to regulate all communication, whether by wire or radio, including telegraph, telephone, broadcasting, satellites, cable, public safety, industry, business, cellular phones, space research, military, shortwave, microwave links, and military uses.

Call Letters

The FCC has the responsibility to approve each station's four-letter call before license approval in order to avoid possible duplicate letter combinations. Except for certain stations that were on the air before the preceding rules were adopted, stations

Band	Uses
VLF (Very Low Frequency) 30 KHz and below	No voice communication Radionavigation, (beacon), maritime
LF (Low Frequency) 30 KHz to 300 KHz	Aeronautical, maritime, radio-location, radionavigation (beacon)
MF (Medium Frequency) 300 KHz to 3000KHz	AM broadcasting, amateur, disaster, industrial, public safety, distress
HF (High Frequency) 3 MHz to 30 MHz	Long-range, international, amateur, citizens band, facsimile, and other
VHF (Very High Frequency) 30 MHz to 300 MHz	Television, FM, space telemetry, satellite, land mobile, maritime, radio astronomy, aeronautical, and other
UHF (Ultra High Frequency) 300 MHz to 3000 MHz	Television, satellite, industrial, space research, citizens band, aviation, public safety, amateur
SHF (Super High Frequency) 3 GHz to 30 GHz	Satellites, television pickup, radar, amateur, space research, radio navigation, and other
EHF (Extreme High Frequency) 30 GHz to 300 GHz	Space research, radio astronomy, radiolocation, experimental, amateur

▶ ***Figure 1*** *The Radio Spectrum.*

west of the Mississippi have *call letters* beginning with K, whereas stations east of the Mississippi have calls starting with W.

In addition, the FCC has the power to license stations, approve renewal or transfer of licenses, and to monitor a station's adherence to good technical standards so the station can serve the most people possible within its signals, yet not interfere with another station's signal.

Radio Waves

Throughout this section keep in mind that the term *radio* includes television and any other service that uses radio waves. Measurement of radio waves is in cycles per second experienced by a wave at a particular frequency. Today they are measured in Hertz (Hz) to honor the man who first generated, distributed, and received radio waves and, thus, was the forerunner of broadcasting. One Hz is equivalent to one cycle per second.

Although the radio spectrum starts at 10,000 cycles per second (10 Kilohertz = 10 KHz) and goes up to 300,000 billion cycles per second (300 Gigahertz), we can use only the lower third of the spectrum since two-thirds of radio spectrum is not practical for communication.

Wavelengths A sound that completes one wave per second is said to have a frequency of one cycle per second (1 Hz). To help visualize a wavelength, it helps to remember that a tone of 1100 Hz has a wavelength of 1 foot, which means that the wave completes 1100 1-foot wide cycles in 1 second. That frequency is close to that of the normal range of the human voice, which has a frequency of approximately 1000 Hz. Near the top of the hearing range of the human ear is the highest note of the violin, which vibrates at a frequency of 16,000 Hz (16 KHz).

The higher the frequency, the shorter the wave, because more waves have to be formed within one second as the frequency increases. Therefore, FM and television, which operate in the VHF range, have shorter and more waves per second than does AM in the medium frequencies (MF).

Carrier Waves Broadcast waves require waves of much greater frequency than the natural sound; these are called carrier waves. For example, if a hypothetical station is assigned by the FCC to a broadcast frequency of 1200 KHz, that station will be found at 120 on the radio dial. But in a sense that station piggybacks on a carrier wave that oscillates 1,200,000 (1200 KHz) vibrations per second. The number 1200 is the center of the station's assigned frequency, which means the station's signal space is actually from 1155 KHz to 1205 KHz. In radio we usually refer to stations by their assigned frequency.

According to the type of assignment, AM stations use power ranging from 250 watts to 50 kilowatts. This power is used to generate the carrier wave.

AM (amplitude modulation) is so named because the broadcast audio modulates (increases and decreases) the amplitude (height) of the carrier wave. FM (frequency modulation), however, modulates the carrier wave's frequency.

FREQUENCY ASSIGNMENTS

Amplitude Modulation (AM)

Amplitude modulation was the first system of broadcast transmission. Each AM station's channel is 10 KHz wide. The FCC has set aside 107 medium frequency (MF) channels for AM of which 60 are classified as clear, 41 as regional, and 6 as local. Within this group, they are further classified as either Class I, II, III, or IV.

Class I operates with high power and exclusively on clear channels. Class II may operate on clear channels but are considered "secondary" because they may operate on a variety of power assignments from 250 up to 50,000 watts, as approved by the FCC. Class III stations broadcast on regional channels and have less coverage than do clear channel stations. Class IV are those stations that operate only on local channels and cover less area. More than 22% of AM stations use one of these six local channels. Because Class IV stations have low power, thus going a shorter distance, a number of them can broadcast on the same frequency, located far enough apart geographically so that they do not interfere with any other station on the same frequency.

More than half of the AM stations are Class I, operating on clear channels, which are the strongest, and were intended originally to serve remote regions. Today there is the possibility that some clears could be broken up to provide spectrum space for *daytimers* (stations that operate only from sunrise to sunset and have to go off the air at night, because otherwise they would interfere with the clear channel stations.)

An AM station has two types of waves: sky waves and ground waves. During the day the sky wave dissipates and is unreliable; therefore, an AM station relies primarily on its ground wave. However, at night the AM sky wave is useful because it bounces back from the ionosphere. Since sky waves can cover great distances, a clear channel station can send its signal as far as 700 miles. For this reason, only clear channels are authorized for sky wave service, to prevent interference between clears and other classes of stations.

Frequency Modulation (FM)

Frequency modulation (FM) propagates signals of greater frequency and smaller waves than does AM, with one channel being 200,000 cycles per second (200 KHz) wide. FM's much wider bandwidth allows the sound to ride in the middle of the carrier waves, thus protecting it from electronic interference that causes static. In addition, FM can handle the full spectrum of audio and, therefore, is preferred for broadcasting music.

Operating in the VHF band, there are 100 FM channels ranging from 88 to 108 megahertz (MHz). Of these the bottom 20 channels are reserved for educational or noncommercial use. A station, for example, found on the FM dial at 91.9 actually has spectrum space reserved from 91.8 to 92 MHz; the number 91.9 reflects the station's center frequency.

FM stations are classified by their power and their antenna height. Class C FMs are the most powerful, with maximum effective radiated power (ERP) of 100 kilowatts (KW). Class B stations are limited to no more than the output of 50 kilowatts. Class C FMs are restricted to no more than 3 KW ERP. Class D is the category of ten watt noncommercial educational stations.

PROPAGATION OF SIGNALS

Role of the Antenna

We have mentioned the importance of frequency, power, class, and the carrier wave. But just as a baseball bat is needed to boost a ball into the air, we need a transmitter for power and an antenna to distribute the broadcast signal into the air.

In the AM service the entire antenna tower acts as the antenna, its height being ½ the wavelength of the station's assigned frequency. Most AM antennas are no more than 1000 feet tall. The AM antenna sends out radio waves in all directions and at all angles, even straight up to the sky. An AM station with a nondirectional pattern uses one antenna. The fewer the obstructions in the area, such as tall buildings, trees, or mountains, the better the coverage.

In AM there are two waves. The ground wave is distributed by underground copper wires radiating from the base of the antenna. The sky waves are radiated by the tower itself. The transmitter provides the initial boost, and the antenna scatters the signal.

Factors which influence coverage include power, frequency, antenna system, time of day, and local conditions, such as climate and soil conductivity. An individual or group considering purchasing a station will depend on a consulting engineer's findings on these matters. It is important to know, however, that greater power does not always assure greater coverage, since stations at lower frequencies often cover a greater area than do those with identical power, but at higher frequencies. Generally speaking, however, the higher the location of the tower, the greater the coverage, especially for daytime AM. Also, only ground waves are strong enough to override static in AM.

AM's sky wave is primarily useful in rural sections since man-made static can interfere with reliable sky wave reception in large cities.

Directional Antennas

Directional antennas consist of more than one antenna, positioned so that they can distribute the signal in noncircular patterns to arrange coverage of inhabited areas without wasting the signal on pastures. This type of coverage is vital not only to station ownership and management but also to the sales department, since the more people the station covers, the higher the rates that can be justified to advertisers.

FM Antennas

In FM and television, transmission is limited to line of sight, and the FM antenna itself is a small unit, no more than about 2 feet in diameter, in varying configurations, mounted on a tower. The higher the tower is above average terrain, the farther the signal will go. Height above average terrain (HAAT) does not necessarily mean height above the ground on which the tower sits. A 1200-foot antenna on top of a 2000-foot mountain would be 3200 feet above *average* terrain.

Television towers may extend above ground as much as 2000 feet but higher towers are more expensive to erect and to maintain. The Federal Aviation Administration also requires special painting and lighting of antennas more than 200 feet high to minimize hazards to aircraft.

TELEVISION

Very High Frequency (VHF)

Television occupies a much wider channel than do the radio services. Each television station's carrier frequency occupies a band width of 6 million Hertz (6 MHz). The picture utilizes approximately 4 MHz, with the remaining portion of the channel used for the FM audio, plus buffer and sideband requirements. We could fit 600 AM stations into the bandwidth of one television channel. Television stations are usually identified according to their assigned channels because it is more convenient to say "channel 2" than to identify the audio and video by their actual carrier wave frequencies.

Because distribution of American television programs by syndication to foreign markets is an important source of income to American broadcasters, the matter of line compatibility in video is critical if a program is to be distributed worldwide by radio waves rather than by film or videotape. In the United States each picture consists of 525 lines per frame, with 30 frames projected every second, based on this country's 60-cycle house current. European countries use a system based on their house current of 50 cycles, and project 25 frames every second with each frame made up of 625 lines.

Ultra High Frequency (UHF)

UHF operates the same as do VHF stations except that UHF, with its higher frequency and resultant narrower channel, requires more power than VHF and does not go as great a distance.

High Definition Television (HDTV)

High Definition Television, a system pioneered by the Japanese, provides 1125 lines per frame and surpasses the quality of theatrical motion pictures. Its technology has begun to replace 35-millimeter film because HDTV produced on videotape is cheaper and easier to edit than is film. Using HDTV in production can reduce costs, with the added advantage of a higher quality picture. Several feature films have already been produced with this method.

Although the Japanese manufacturers expect to begin HDTV broadcasts in Japan by 1991 with a system they call MUSE, American broadcasters are waiting for the FCC to establish technical standards and possibly issue licenses for HDTV broadcast licenses in the United States.

The challenge is to transmit HDTV signals so that they can be received on conventional television sets, a problem similar to the compatibility between color and black and white when color was introduced. HDTV requires 30 MHz for one channel, which is five times as much spectrum space as one of our conventional television channels. One proposal has been to compress the television HDTV signal with digital components so that all currently operating regular television stations could be provided an additional 6 MHz for HDTV transmission. Another suggestion has been to bypass the conventional station and transmit by direct broadcast satellite (DBS). A third proposal has been to use cable's optical fiber links to replace present coaxial systems that could not transmit wide-band television without serious signal degradation. Some recommend use of multipoint distribution systems (MDS) which utilize microwaves.

The FCC's Office of Engineering and Technology, responsible for testing and approving systems, is seeking a method which could use existing channel widths and be receivable on present home television receivers. We must keep in mind, however, that for HDTV's quality to be appreciated, it is best viewed on a larger screen. The economic question is whether Americans would be willing to invest in expensive new receivers for such reception.

Low-Power Television (LPTV)

Low-power television stations (LPTV) currently serve two types of roles. Some continue their original function as *translator* stations, which pick up the original channel and then rebroadcast the signal on a channel with a different number in order to provide a better signal to metropolitan areas not otherwise served adequately because of poor reception of the original signal.

Other LPTVs function exactly as do more powerful stations except that the coverage area is smaller. Programs are usually geared to special audiences, to offer programming not of interest to the general mass audience. Some LPTVs serve ethnic groups. And today in some isolated communities LPTV stations may be the only source of local news.

Occasionally LPTV networks are formed and are distributed by cable or satellite. For example, Channel America's 24-hour-a-day LPTV network consists of seven 0 & 0 LPTV stations plus 12 affiliates.

As the name implies, LPTV operates at much lower power than do "traditional" stations, and uses only 10-watt maximum power if on VHF and 1000 watts if on UHF. Coverage ranges from 15 to 25 miles contrasted to the 40-70 mile range of full power television stations.

CABLE

Cable and cable systems have come a long way since they were inaugurated as a way to provide television to homes situated in areas where terrain prevented television signal reception. Because it originally consisted of an antenna perched on a high spot from which cables went down to home receivers, and did not utilize regular on-air transmission, cable was defined by the FCC as a *common carrier*, just as is telephone. By 1990 there were some 9300 operating cable systems in the United States, serving some 25,000 communities, reaching more than 47 million subscribers (perhaps 130 million people), with 53.8% of the nation's television households wired in (*Broadcasting/Cable Yearbook*: 1989). The largest multiple system operator (MSO) was Tele-Communications with over 9.5 million subscribers.

Cable systems are franchised rather than licensed and are regulated by the municipal governments from which they need permission to run cables over or under public property. While some have single owners, many have multiple owners. Regulatory aspects are discussed in Chapter 6, "Regulation and Controls." As technology continues to become more sophisticated, cable systems are providing additional services beyond basic cable. These services include pay cable, automated origination, access channels to be programmed by persons other than the cable operator, plus two-way capability between a subscriber and a system.

SATELLITES

Satellites may be both allies and competitors of cable and broadcasting, depending upon who uses satellites and to what economic advantage.

Technically, communications satellites are launched into positions 22,300 feet above the equator where they orbit around the earth at the same rate as the earth's rotation, thus appearing to be stationary. Although the first commercial satellite, Intelsat, was put into orbit April 2, 1965, it was not until 1975, when Home Box Office (HBO) began to transmit Pay-TV programming over Satcom I, that commercial interest in satellite use began to grow. In 1990 there were 26 satellites in orbit, of which 3 have transponders dedicated to cable program distribution. Public Broadcasting Service and commercial broadcasting use satellites for network program distribution.

Either C-band or Ku-band frequencies are used by an uplink earth station to feed the signal to the satellite. C-band equipment is relatively inexpensive and has sturdy technology, but it requires careful coordination of both the uplink and downlink sides with terrestrial microwave facilities that may share the same frequency. Ku-band equipment is more sensitive, but it lacks some of the advantages of C-band. But Ku-band technology is preferred because it uses a much smaller receiving dish, which may be only 3 feet in diameter, compared to C-band receiving dishes with a diameter of at least 35 feet.

Direct Broadcast Satellites (DBS)

At this writing the FCC is encouraging development of DBS because individual homes could afford and accommodate the smaller receiving dish. DBS could expand the home satellite market several times more than the current two million consumers presently receiving programs on C-band. The 100th Congress passed copyright legislation in 1989 with the hope that it would help the developing home satellite industry by authorizing transmission of broadcast television signals via satellite to backyard dish owners.

Under terms of the Satellite Home Viewers Act of 1988, the signals of independent television stations are allowed to be beamed to homes with dishes. However, signals of network affiliates may only be delivered by satellite to those areas not able to receive network programming off the air in the so-called "white" areas.

A consortium called Sky Cable is studying new technology that, possibly by 1993, might enable people to clip satellite dishes the size of laptop computers outside their windows. These flat, 12-inch square antennas will be able to pick up satellite signals because satellites of the future will have much more power than those provided in 1990. With the anticipated new 200-watt transmitters to feed it, the receiving dish need not be as large as those currently in use.

Satellite Master Antenna Systems (SMATV)

Although it uses cable-type technology, the significant difference between cable and SMATV is that SMATV systems operate only in private properties such as condominium complexes, private housing developments, apartment buildings, and mobile home parks, all of which pick up their programs by hooking either to an earth station aimed at a cable satellite or by using a master antenna system wired into the building for that purpose.

From management's view, SMATV is operated technically and administratively the same way as cable. The only differences are that SMATV systems are free from regulation because they do not require a municipal franchise and also are generally excluded from the FCC's definition of a cable system.

MICROWAVE SYSTEMS

Instructional Television Fixed Service (ITFS)

Utilizing microwave signals to transmit programs, ITFS consists of a block of channels originally reserved exclusively by the FCC for instructional, cultural, or other educational programming. Not all the channels are in use. The majority of channels are used by the Catholic Church to originate and feed instructional material to their own schools.

Multipoint Distribution Service (MDS)

In 1983 the FCC reallocated eight ITFS channels to multipoint distribution service (MDS), creating what may be called *wireless cable*. MDS uses omnidirectional microwave signals and can transmit video, data, text, and other services to subscribers, who must be equipped with special antennas to receive the signals. Most MDS time is leased to pay-movie entrepreneurs, who provide programming to hotels, apartment buildings, and homes. Time will tell whether MDS will become a significant factor on the pay television scene, and whether it can preempt or compete with conventional cable.

SUPPORTIVE TECHNOLOGIES

Fiber Optics/TELCOS

All of the preceding methods of transmitting television signals had one problem: they were security risks when one wished to send matters of top security. Then came fiber optic transmission, which provides the necessary security.

A fiber optic cable is approximately the thickness of a human hair. Its transmission depends upon a precision glass cable which transmits communication information by using an infrared light beam, providing a high quality television signal that would be extremely difficult to intercept or "tap."

Historically, the first networks could not have existed without the telephone companies (TELCOS). Now the TELCOS are interested in becoming information providers, a move seen by broadcasters as the death of free TV. In addition, cable companies fear the advent of fiber optic distribution by the TELCOS; even the American Newspaper Publishers Association is concerned about the possibility of fiber optics entering electronic publishing.

At present the TELCOS are classified as a common carrier, allowed to provide information-carrying *services* to all, but forbidden to originate information. Because fiber optic technology can provide superior audio and video services with the same ease as they carry data, a telephone company using fiber optics and already serving most homes could overwhelm free broadcast television. In addition, through use of fiber optics the TELCOS could not only offer interactive services and pay-per-view movies but could even deliver HDTV.

This was an important item on the agenda at the FCC in 1990, and action was expected to take place about the matter as to whether there should be a ban against telephone companies owning cable systems in their telephone service areas. It is possible that TELCOS will be allowed to serve as transporters of others' programming, but not to be programmers themselves in their own service areas.[1]

Computers

The role of computers has expanded beyond aspects of operation and production. Computers assist scriptwriters by automatically printing the correct formats as the author types a script. They are also being introduced at the Supreme Court to distribute the court's opinions within minutes of pronouncements from the bench, replacing reporters having to spend hours waiting for summary "handouts" about decisions.

Computer as a Management Tool The computer is a valuable management tool. Although business simulation has existed for years in corporate rooms and graduate-school labs, only in the late 1980s did powerful personal computers and simpler software take technology to average managers. With computer simulation a manager is able to work ahead, instead of living from crisis to crisis. Simulations are also being developed for employees to enable them to see the big picture, whether it is in product development, sales, projected audiences, or in designing engineering facilities. With computer simulation, management can experiment with "what if" without taking unexplored risks.

Computers also handle billing, proof-of-performance affidavit, logging, and scheduling daily program logs. The computer enables television studios to use automated consoles and cameras, thus eliminating some human responsibilities. Radio stations can now transmit computer programs on their subcarriers as well as their main channels. Moreover, the computer is also used in some systems for teleconferencing via satellite.

Other Technologies

The 1980s was a decade of high-tech electronics. In addition to those already discussed, other technologies are of concern, directly or indirectly, to broadcasting, cable, and to media managers. We have video recorders, compact discs, AM and FM stereo, digital paper, and cellular radio. Some of these are supportive, others may be either supportive or competitive in matters of program distribution. Each has an impact upon cost-effective management.

SUMMARY

The 1990s are likely to see some major changes in the ways people receive their electronic entertainment. To make up for the lower profits created by the competition, many broadcasters are already buying into some forms of alternate technology, especially cable at this time, but also possibly satellites and fiber optics. The primary concern of the FCC is to assure that there will be some form of free system available to the economically disadvantaged citizens, who rely solely on broadcasting for their information and relaxation. It is thought that for this reason the FCC may ease regulations of the networks, since networks seem the logical source of the free product. Perhaps if networks are permitted to own and sell the syndicated material, as discussed in Chapter 6, the revenues thus gained would support free broadcasting. Obviously this will be a decade of change and challenges. Technology is the tool. But programming and economics will determine how that tool is used.

Notes

1. *Broadcasting*: 15 January 1990, p. 120.

3 Media Ownership

BROADCAST OWNERSHIP

Successful media managers must know not only the workings of the particular medium for which they are responsible, but also must understand the American system, which generally differs from most foreign systems in matters of ownership, financial support, and control. Management and staff must be alert to ownership's viewpoint, philosophy, and objectives. This chapter reviews media management as it relates to ownership in the United States.

In addition to broadcasting, *electronic media* include those which use some forms of electronic distribution and include cable, satellites, microwave applications, telephones, and fiber optics. We also find that conglomerate or group ownership may include a mix of newspapers, magazines, books, and entertainment industries such as movie studios and record firms.

Media in the United States are all privately owned, in contrast to other nations in which all or some media are government-owned and controlled. Through the Communications Act of 1934 Congress declared that the "airwaves belong to the people." Because there is no financial support from government, America's radio and television industries developed as privately-owned, advertiser-supported media.

Educational broadcasting, later renamed Public Broadcasting, is funded by non-profit organizations, groups, and listeners.

The distinctive factor of broadcasting in the United States is that the general public receives the programs at no expense to themselves. Other electronic media charge for their services in some manner.

The broadcast media are regulated by the government, through the FCC, but this does not imply control such as that found in other countries. Regulatory matters exist to provide the best technological services, to prevent stations from interfering with one another, and to protect First Amendment rights.

Although the U.S. government does not own facilities that broadcast directly to the public, it does own systems and uses portions of the spectrum needed for such services as air traffic control, weather surveillance, satellites, and the armed forces.

Qualifications

Because broadcast licensees use the public's airwaves and are expected to serve primarily U.S. citizens, they must hold U.S. citizenship. In addition, the character of prospective applicants is considered; for example, persons convicted of serious crimes may not be granted licenses. Applicants for a broadcast license must also show the FCC that they are legally, technically, and financially qualified and that their proposed operation would be in the public interest. Since deregulation in the 1980s, the FCC places less emphasis on financial or programming aspects than formerly.

Ownership Opportunities

Opportunities to own radio or television stations are confined to a choice between existing stations up for sale or finding an available unused, appropriate frequency in the market to be covered. One usually engages a broker who specializes in such matters to help find a prospective property.

It is important to remember that ownership applies only to physical properties such as buildings, equipment, and ground. A license is never *owned*, and, therefore, can never be sold. Besides, a licensee never owns the frequency on which the station broadcasts. Prospective owners must file with the FCC to apply for a specific license for a particular station or area, a process which is described later in this chapter.

Existing stations are sold either because the owner has chosen to focus upon some other type of investment, or is trading up to a larger, more potentially successful operation. Prices are determined by the financial success of the station, and that depends upon how many homes the signal covers, the strength of the signal, and strong ratings. In 1989 the price of stand-alone TV stations averaged $18 million; the average stand-alone AM cost an average of $745,000, while the average FM stand-alone price was $1.6 million. An AM-FM combination brought an average of $3 million.

If there is competition for a particular frequency, the FCC favors local ownership when possible, based on the assumption that local owners are more familiar with their communities and, thus, are able to anticipate and provide good community service.

Corporations unable to be licensed include those with foreign officers or directors, or those in which more than 20% of the capital stock is controlled by foreign interests.

Buying a Station

Price Determination Station prices depend on a number of factors: demonstrated economic success, cost of facilities, market size, network affiliation (if any), age of station, type of ownership, and whether the facility is AM, FM, VHF, or UHF. The most important are the market size and the number of homes reached. Media brokers and financial analysts help determine the market price and, for their services, receive a commission as a percentage of the selling price. Media financial analysts calculate existing sales and potential profit. The best predictor of the sales price of a television station is based on the audience size and its average prime-time household rating.

Originally AM was the preferred radio medium, with FM tagged on, often merely to duplicate AM's programming. FM gained listeners later as a classical music outlet because of its superior sound. But not until the 1980s did FM become as valuable as an AM property; and by 1989 the average price of AM stations had dropped about 4% a year while FM stations jumped an average of 20% in value. One reason, apparently, is FM's superior stereo compared to that found in AM home stereo receivers. Radio station prices are estimated on the cost per potential listener. In cable it is the cost per subscriber that establishes the price.

Financing Where does money come from to finance building or buying a station? Money sought to build a facility is referred to as venture capital, whereas money raised to purchase an existing facility is called acquisition financing.

Internal capital is raised by combining the assets of a proposed partnership, joint venture, or corporation. If not enough internal capital can be raised, then external capital, called venture capital, is raised through outside sources such as government bonds and loans, banks, insurance companies, pension funds, commercial credit firms, small business investment companies, as well as venture capital firms.

In the 1980s leveraged buyouts were used more frequently at the upper level but are more risky since they use borrowed money to finance most or all of the purchase price. However, that trend began to fade in 1989 as the station-trading boom began to slow down. At the beginning of 1990, there were numerous companies that put their properties on the market to bring down their debt load or to divest to go into other types of investment which put a lot of station inventory up for sale in the 26th to 100th markets. However, if a property is in a desirable location and more than one person is interested, brokers say that the selling price should be every bit as strong as it was at the peak of the 1980s.

Useful for the potential buyer is the Broadcast Investment Analysts' booklet, *Investing in Television*, which contains summary information for more than 200 television markets on rating trends, station sales, population projections, and advertising expenditures.

Joint Ventures The 1980s introduced a period of cooperation between former media rivals. Broadcast companies, for example, launched program ventures with cable television operators; newspaper and broadcast companies joined to study uses and application of teletext. Joint ventures occur among small media markets as well. Local broadcasters have joined with newspapers, and even the media giants and networks invest in diverse activities such as satellite channels or special programs for particular types of audiences.

In a sense, the cooperation between large print companies and entertainment groups, as well as the cooperation between broadcasters and cablecasters, began to rise in 1989: a big switch from the 1970s when broadcasters considered other groups and cable entrepreneurs to be mortal enemies.[2] It takes much longer to start a new radio or television station than it does to buy one, but even the latter process also may take more than a year because of the backlog of processing applications at the FCC. Available spectrum space for broadcasting is extremely limited, and the search for an available frequency is time-consuming.

Occasionally a station may become available because the station's license was not renewed. However, denial of renewal is rare and stations with a satisfactory record should expect no problems at that time. Limited opportunities, therefore, make it more difficult for a newcomer to break into ownership.

It requires a very high initial investment to construct or acquire property in broadcasting and cable. Radio construction costs range from $50,000 for a small market AM--FM up to several million dollars for a sophisticated radio facility in a major market.

It is easier to get into low-power TV (LPTV). A LPTV with production capability may cost no more than $300,000 compared to a full-power television facility that may cost $3 million.

Ownership by Women and Minorities The chain of events initiated by the Civil Rights Movement and the subsequent Kerner Report led the FCC to encourage minority ownership, believing that participation by women and minorities[3] beyond the clerical or disc jockey level would be in the public interest for two reasons: (1) it would increase diversification of control of media and (2) it could also provide more program variety.

Historically, the tradition of white male ownership of broadcasting began in the first decade during which broadcasting was primarily the province of white engineers who experimented, developed, and improved broadcasting equipment. Quite often they then moved into ownership. Not until almost 50 years later did ethnic minorities, and later women, begin to enter the higher levels of broadcast ownership in the 1970s. Even then, progress was slow because of funding problems as well as the scarcity of available channels or stations. To boost minority ownership, Mark Fowler, then FCC chairman, promoted a management training center for minorities and women, established procedures giving preferential treatment to women and minorities for ownership of stations that had lost their licenses at renewal time, and the concept of a reformed tax code. In the 80s minorities continued to achieve some ownership, but funding was still the major hurdle. The FCC continues to seek new money and regulations to help increase minority ownership, particularly of television stations.

Minority ownership remained low by 1990, with the number of minority-owned commercial stations being only 300 of the 11,500 commercial stations on the air. Pluria Marshall, head of the National Black Media Coalition, credits FCC policies for about 150 of those. In Marshall's opinion the most helpful FCC policy has been issuance of tax certificates to broadcasters who sell their stations to minorities. He reports that the FCC's distress-sale policy has not been used more than about ten times since 1980.[4] James Winston, executive director of the National Association of Black-Owned Broadcasters (NABOB) supports the FCC policies even though he believes they have not been particularly effective to increase minority ownership. Winston says that "without them it will be devastating. We'll be closed out of the ball game."[5]

Most minority ownership is in radio, with blacks owning approximately 260 radio stations and Hispanics about 28. Only 11 television stations are minority-owned. Regan Henry, the first major black entrepreneur to own broadcasting properties, established a record for station trading in 1989 when he signed letters of intent to purchases 10 AMs and 13 FMs.[6]

Although the percentage of women who participate in ownership looks higher on paper, they are generally stock holders and are not active in the field. Some women, however, do own small radio stations or share ownership as in mom-and-pop enterprises.

Ownership Limitations

The FCC permits a single entity, whether an individual or a group, to own a maximum of 36 stations: 12 AM, 12 FM, and 12 television, provided that the television stations collectively serve less than 25% of the nation's television homes. Exceptions apply to UHF properties. Those stations count for only half of a market's television homes. In addition, if minorities control 2 each AM, FM and TV stations, then groups may own as many as 14 stations of each type and blanket up to 30% of the nation's television homes.

In order to avoid a local monopoly of information and entertainment as well as to stimulate diversity, the FCC has one-to-a-market rules, which permit a single entity to own an AM-FM combination in the same market, but not to own a radio and a television station in the same market. Moreover, no owner may operate more than one station of each type in the same service area.

Cross ownership rules also affect the number of stations that may operate in certain markets. Although a station may be co-owned with other media, the FCC's *Newspaper-Broadcast Cross Ownership Rule* forbids ownership of broadcast stations and newspapers in the same market. In a similar sense, the *TV-Cable Cross Ownership Rule* forbids television stations to be co-owned with cable television systems in the same market.

Because television has been losing some of its viewers to other technologies, there is the possibility that the FCC may rescind some of its rigid regulations so broadcasting can be economically competitive.

Licensing

Obtaining a License In broadcasting, the potential owner requires the services of outside consultants such as brokers, communication lawyers, and engineers to expedite the licensing process. Full details of station licensing procedure are found in Part 1 of the FCC's rules, "Practice and Procedure," and also in Part 73 of "Radio Broadcasting Services." Copies of the complete rules may be purchased from the Government Printing Office. The procedure is basically the same for seeking licenses for all three types of facilities, AM, FM, or TV.

The names of firms and individuals practicing before the FCC are listed in various trade publications. They may also be identified through the state broadcasting associations.

Applicants are responsible for their own search to find a frequency on which they can operate without causing or receiving interference from existing stations, or those with pending license applications. In addition, applicants must request the desired community, power, channel, and class of station.

After license application data is gathered, the next step is to apply for a construction permit (Form 301), which requires information about the applicant's citizenship, character, finances, type of transmitting apparatus to be used, antenna and studio locations, and the proposed service.

Within 18 months for television and 12 for radio, the recipient of the construction permit (CP) is expected to construct and test the station. Then, with the appropriate form from the FCC (often a telegram), the potential licensee may begin to test programs on the air while waiting for the license to be awarded.

In the case of network owned-and-operated stations (O & Os), the FCC licenses only the individual stations, not the network.

Keeping a License The maximum term of a radio station license is 7 years; however, the maximum term for a television station license is 5 years. At the end of each license period, station owners apply for a license renewal, at which time it is possible that competing applications by other potential owners may be presented. The most severe penalties which can affect licensees are short-term renewals or revocations of license.

A majority of license renewals are uncontested if there have been no serious complaints, penalties, or opposition filed. Licenses have been withdrawn or renewal denied if broadcasters have been proved to have deliberately deceived competitors, clients, advertisers, or the FCC.

License Challengers Parties who oppose incumbent licensees but do not want to take over the license may file "petitions to deny renewal," usually because they feel that incumbents have failed to meet public-interest standards. Should the petition succeed, the station may then be available for sale to an entity that believes it can serve the public interest more equitably.

In the case where would-be licensees believe they can serve the public interest better than the incumbent, the FCC holds a *comparative hearing* before an administrative law judge. The decision is based on criteria designed to promote diversity and integration of ownership and management. Incumbents try to show they have provided "meritorious" service by supplying the administrative law judge with programming schedules and testimonies to their good citizenship. Simultaneously, the challenger seeks to rebut the list of community programs (especially if such were not aired), or they may claim that the aired programs were "not responsive" to issues of local concern.

The process, obviously, is time-consuming and expensive. At this writing many broadcasters have been asking Congress to modify the system so that competing applications would not be considered until *after* the FCC decided not to renew a incumbent's license. It is a matter likely to be debated for a long time, but an essential discussion in a democracy concerned with serving the public via limited facilities.[7]

TYPES OF OWNERS

Commercial Broadcasting

There are several categories of owners, as defined by the FCC. *Single* ownership occurs in the smaller stations, usually a mom-and-pop operation. *Local group* ownership may occur when members of one family or several local people form a corporation for the purpose. Mom-and-pop stations generally are low power and often a daytimer only.

Group ownership is defined as individuals and companies with multiple station holdings and includes all which control three or more radio stations and/or two or more television stations. The usual 12-12-12 limits, discussed earlier, apply. Statistics reveal that there is a national trend toward group ownership of broadcasting properties and, when measured by market size, more than 80% of the television stations in the country's top 50 markets are group-owned.

The majority of broadcast stations and cable companies are owned by *corporations* for safety as well as financial reasons. A corporation is considered as one person when making contacts, paying taxes, and settling debts; thus legally the members of corporations are more protected than are single owners of enterprises.

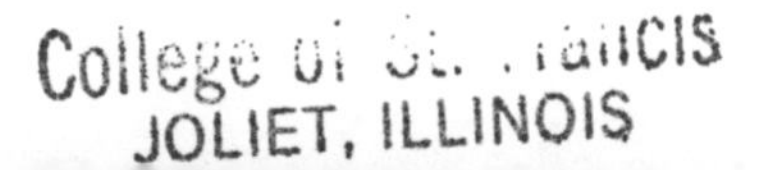

Networks There are more than 500 commonly owned station groups, most of which own three to five small stations. However, the three major networks' owned and operated (O & O) stations are very powerful groups. To insure compliance with the FCC requirement that each station must serve its own community of license, each O & O is administered separately from network operations, and has its own general manager with considerable freedom in operating the station. There are also a number of large non-network group owners.

Network ownership changed hands in 1985. American Broadcasting Companies, parent of the ABC network, was taken over in 1985 by Capital Cities Communications and is now named Capital Cities/ABC Inc. What is noteworthy is that Capital Cities was a smaller operation than ABC. It was the result of a so-called leveraged buy-out. To comply with FCC rules that no one entity can reach more than 25% of the nation, Capital Cities sold its MSO and also had to divest itself of more than $1 billion of print operations.

General Electric (GE) purchased RCA in December of 1985. RCA was the oldest and largest of broadcasting companies as well as the parent company of NBC. Thus NBC became part of the nation's second largest industrial corporation.

Conglomerates A conglomerate is a company that operates in a number of different and unrelated markets and is different from a corporation, whose holdings may be varied but are all related to the communications field. A conglomerate, on the other hand, may have non-media ownings in addition to broadcasting holdings. For example, at one time RCA owned not only NBC radio and television networks but also owned real estate, a frozen food corporation, and a carpet manufacturing firm. In today's atmosphere of corporate takeovers, such conglomerate holdings change rapidly. Some concerned media critics are wary of conglomerates, fearing that concentrated media ownership offers the potential to use such media to advance the interests of their non-broadcast holdings.

Noncommercial Broadcasting

The structure and ownership of noncommercial stations differs from that of commercial organizations. In public broadcasting, licenses are held by four different types of entities: (1) states or municipalities, (2) college and universities, (3) public school systems, or (4) communities.

College and university stations were the first to go on the air, offering courses and programs on radio since the 1920s. Today both public radio and public television stations licensed to educational institutions continue to provide these services.

Some states have statewide networks for which one flagship station provides the programming and study guides, and the other stations transmit the programs to schools around the state, especially Alabama, Nebraska, South Carolina, Georgia, Kentucky, and Mississippi. In Pennsylvania the network is informal in that the stations sharing programs are actually licensed to different groups, such as educational institutions and local communities.

Local school boards use t heir stations primarily for in-school instructional programs. Some community groups also own public television stations.

The Public Broadcasting System (PBS) was established in 1967 by the Corporation for Public Broadcasting (CPB) to arrange network facilities to link noncommercial television stations nationally. In 1978 it was PBS which pioneered use of satellites to distribute programs to member stations in place of land lines.

Cable Ownership

There are three general types of cable ownership. The smallest unit is the mom-and-pop system. Above that comes a larger, independent cable system. And finally there is Multiple System Ownership (MSO) in which a MSO may buy an unlimited number of systems. There are over 180 MSOs in the United States. Some MSOs are owned by broadcasters, others by television producers and distributors. Magazine and newspaper publishers such as *Time, Inc.* are MSOs, as are conglomerates. Thus we find cable systems having ties with broadcast interests, program producers, and publishers. At present there are no restrictions on the number of cable systems an entity can own, but the FCC has occasionally considered adopting formal guidelines related to group ownership of cable television firms.

Although ownership of cable systems differs from that of broadcasting, it is still a matter that directly involves local as well as national regulations and procedures. Both broadcasters and cable operators own and operate their own facilities: both are regulated, however, in certain matters. As pointed out earlier, the FCC regulates broadcasting solely at the federal level. Although cable is regulated federally in some aspects, franchising a cable system (comparable to licensing a broadcast station) is strictly a local matter, because the FCC realized that federal licensing of cable would be awkward due to the number and variety of local communities to be served. For this reason, the FCC adopted minimum standards for franchises to be issued by *local* governments in matters concerning the franchising process, authorization to access to public right-of-way, construction deadlines, and establishment of a franchising authority office at the local level to administer the franchise and to handle complaints.

Obtaining a Franchise The FCC prescribes rules concerning ownership or control of cable systems by persons or groups owning or controlling other media of mass communication witch serve the same community served by the cable system. But the FCC does not license cable.

A would-be owner, called an operator, must seek a franchise through a local community governing body, such as a city council, community board, or civic or neighborhood associations. Only a few local governments operate their own systems. Generally, a typical cable system operator is private and corporate; it tends to be a multiple system operator (MSO). The local franchising authority handles cable license renewals. The authority need not judge on the public interest standard to renew a cable license. However, if it wishes to deny renewal, the authority must hold a hearing to determine whether the incumbent has complied with the law, provided the promised service in light of community needs, and has maintained the financial, legal, and technical requirements.

The franchising process begins when a local governing body decides it wants a cable service. First it draws up a formal legal *ordinance* which describes conditions under which a system will be permitted to operate. The ordinance usually establishes the term of the franchise (normally 10–15 years), the quality of desired service, technical standards (including the minimum number of channels and interconnection with other systems), and the franchise fee. Franchises are awarded on a competitive basis to a successful bidder. A few communities operate cable systems themselves. It is also possible for a franchise authority to grant one or more franchises in its territory.

The potential cable operator should know the community, its leaders, and the local political structure, as well as the type of available television reception in the area. The latter is important because some communities may be less interested in the expensive extras proposed by many franchise bidders and most interested in cable primarily for better off-air reception.

Obtaining a cable franchise has become a highly competitive matter. With the smaller cable systems being bought out by MSOs, the local governments see increasing numbers of companies competing for local franchises. Some cable operators engage consultants to prepare their franchise proposals. Large corporations may hire full-time staff members to work solely with franchising matters. On the other side, communities also often hire their own consultants to evaluate franchise proposals. Occasionally some systems promise more than they can deliver, necessitating that the community cancel that franchise and go through the entire process again.

Although most franchise contracts last 10 to 15 years, an effective operator should begin to plan for refranchising as soon as the original franchise is obtained. It is easier and more efficient to keep a running assessment of community program needs and desires. Good operators also recognize that their success depends on a staff willing to work with the public so that customers feel their interests and needs are of primary concern to the operator.

Franchising details are spelled out in the Cable Act of 1984 as also are ownership restrictions. Whereas that legislation initiated a trend toward less regulation on the national level with the intent to permit cable to expand as the marketplace developed, in 1990 these were indications that Congress might reregulate cable in some matters as a dozen reregulation bills were pending in Congress at that time.

Cable Revenue Sources Cable's revenue sources differ from those of broadcasting. Whereas broadcasting earns most of its revenue from the sale of advertising, cable's revenue is generated primarily by its subscriber base. Cable systems derive most of their gross revenue from subscriber fees, and earn less than 5% from advertising.

Cable's total potential audience is referred to as *homes passed* and represents the maximum possible subscribers. That number is used to estimate the limit of income available from subscriber fees. The franchise authority's revenue comes from a fee based on basic services only, at a rate of no more than five percent of the franchisee's gross annual revenue from basic services. The authority receives no income from cable's *add on fees*, such as pay-TV and special services.

There are four classes of subscribers. The largest consists of *basic subscribers*, who pay a monthly fee for basic cable service, which entitles them to local and regional television channels, a community bulletin board, satellite-delivered superstations, and advertiser-supported cable networks.

At the next level are *premium* households, which pay an additional fee to subscribe to a pay service that entitles them to such services as HBO and Showtime.

The *multitiered household*, the third level, is that which subscribes to more than one pay service. The top level is the *pay-per-view* household, which pays additional fees for events such as special sporting attractions, first-run films, and live musical or theatrical performances.

Cable Services There are three types of cable services: (1) basic cable networks, (2) pay cable networks, and (3) superstations. *Basic cable* systems generally provide the basic cable network programs as part of the regular monthly subscription fee. Since many of these are advertiser-supported, they cost the cable system very little. A few cable programming services pay the cable operator to use their programming. And each system carrying programs such as the Home Shopping Network receives a percentage of all sales generated by the shopping network's programs on the cable system.

Pay Cable To receive pay cable, subscribers are charged a monthly fee on top of that paid for basic cable service. HBO, the first pay cable network, dominated pay cable for a long time with its large holdings of hundreds of rights to old and new Hollywood features and short films. Even today HBO is said to practically dictate film prices to Hollywood producers. HBO gained its prominence when it initiated the first national satellite interconnected pay network in 1975 by using transponders' time leased on the Satcom satellite.

Cable operators not only may contract for packaged pay programs but can also lease a channel to a pay program operator. They may also obtain their own programming directly from a supplier. Among noteworthy pay cable program suppliers are HBO, Showtime, The Movie Channel, Cinemax, The Playboy Channel, The Disney Channel, BRAVO, American Movie Classics, as well as pay-per-view general programming service. By 1990 pay cable was on approximately 7400 of the 9300 systems in the country, reaching 35 million subscribers in 50 states with approximately 82% penetration of their subscriber count.

Superstations Originated by Ted Turner, superstations are independent television stations whose signals are distributed by satellite to cable systems throughout the country. The cable systems which buy superstations' satellite feeds pay a few cents per subscriber per months to support the satellite uplink and use of the satellite's transponder. Superstation operators make their profit through charging higher advertising rates on the system, justifiable by the larger audience of cable subscribers that can be reached.

The major superstations include WTBS, WBN, WOR, WPIX, and WTVT. Some cable systems also carry the signals of radio superstations, including the classical music station WFMT of Chicago. Cable systems which carry superstations must also pay a fee to the federal Copyright Royalty Tribunal for carrying the copyrighted material in geographical areas not covered by the local superstation's market.

Cable Expenses The largest percentage of cable's budget is spent for programming. Some 60% of revenue is needed to pay for special program services. In the 1990s cable systems also pay most advertiser-supported networks, a reversal of the 80s when such networks as ESPN, WTBC, and others paid the cable systems to carry their programs.

Program expenses escalated by the 90s, not only for direct program purchases and services but also for copyright fees. Prior to 1976, the copyright situation was unclear. After passage of the new copyright law in 1976, cable TV began to pay royalties for transmission of copyrighted works in the form of compulsory licenses paid to the registrar of copyrights, who distributes appropriate funds to copyright owners. The same law established a Copyright Royalty Tribunal (CRT) to resolve disputes, to distribute royalty fees, and to review the fee schedule every five years. It is expected that anticipated changes in the must-carry rules will also increase royalties that cable operators will have to pay.

Cable's second largest expense category is for general and administrative (G & A) costs, approximately 25%. Cable's G & A expenses tend to be higher than those of broadcasting because cable must employ larger secretarial and clerical staffs and needs more expensive computer equipment for billing, handling connection and disconnection orders, and other operational matters of no concern to broadcasting.

Approximately 10% of cable's budget is spent on marketing and includes costs of advertising and promotion, plus the salaries and expenses of sales personnel. Marketing expenses for cable may rise to 25% in the 90s.

Cable's Economic Challenges The 1990s presented cable with a number of concerns. Although industry leaders predict that pay-per-view may be able to provide as many as 100 monthly events, others fear that pay-per-view growth may be hampered through inept scheduling and marketing by some cable operators.

The ratio of basic to pay, as well as to multi-pay subscribers, began to decline in the late 80s. Affecting the ratio were video recorders, competition from retail video stores, which rented major films for as little as $1, as well as the redundancy of premium services.

New alternate technologies also threatened to cut into cable industry's growth. Direct broadcast satellites (DBS), with the capability of providing 100 channels by means of a small, window satellite dish, may woo some subscribers from cable. An additional threat to cable comes from a lobbying campaign by telephone companies which seek to enter the cable business and could probably undercut cable operators' fees, putting many out of business. Wireless cable is also expected to cut into cable's revenues.

Adding to cable's worries in the 90s is the possible return to some form of government regulation. Cable's income would be weakened if broadcasters' efforts are successful to require that cable pay for retransmission of broadcast signals. Also of concern to cable is broadcasters' desire to require that cable locate broadcast signals on the bottom of the dial (a preferred location because "channel flippers" tend to start at the bottom of the dial). Last, but not least, is the expectation that in the 90s Congress might put a cap on the monthly fee which subscribers must pay for basic cable services. After Congress deregulated the cable industry in 1984 and allowed cable to charge whatever the market would bear, the monthly cable rates rose steadily. Placement of a cap on rates would cut into cable's ability to buy programming, to expand systems, and to develop new programs.

Professional Cable Associations Almost all MSOs belong to the National Cable Television Association (NCTA) as do approximately half of the nation's cable systems. It was through NCTA's lobbying that much previous federal deregulation of cable was accomplished. NCTA also provides special services to subscribing cable systems, including management and financial guidance, promotion concepts, and research services. Other associations include the Cable Television Administration and

Marketing Society (CTAM), Women in Cable, and the National Federation of Local Cable Programmers (NFLCP).

SUMMARY

This chapter provides an overview of matters related to broadcasting and cable ownership, including the role of the FCC, ownership opportunities, and types of ownership, as well as how to obtain a license and to finance building or buying a facility.

We have seen that to buy or build a radio or television station is time-consuming because of the search for a facility and the licensing process. To buy a station requires the services of a broker that specializes in selling stations. Because of the scarcity of appropriate spectrum, the opportunities to build a new station are extremely limited. To build a station involves the services of a communications engineer and also a communications law firm. If there is competition for a particular frequency, the FCC favors local ownership when possible.

Compared to the total population, there are few minorities or women who own stations. To encourage such ownership in order to expand diversity of views, the FCC has established some preferential processes and funding sources.

The FCC limits the number of stations that any one entity may own. Classes of owners include: single, group, single with an AM-FM combination, cross media, network, and conglomerate.

The price of a station depends upon the market size, number of advertisers, age of the station, type of facility, and presence or absence of network affiliation. Money raised to build a station is called venture capital; money obtained to purchase an existing facility is acquisition financing. It requires a very high initial investment to construct or acquire property in broadcasting and cable. Less costly and more available television ownership opportunities are found in low-power television (LPTV).

The FCC does not license cable systems; but a potential cable operator must obtain a franchise from a local community-governing body. Although there are no restrictions as to the number of cable systems an entity may own, the FCC has occasionally considered adopting formal guidelines concerning group ownership of cable television firms. Group owners in cable operate in a manner similar to broadcasting, but under different titles.

Noncommercial public broadcasting may be owned by a number of different entities: municipalities, states, colleges, universities, public school systems, or local communities.

Notes

1. *Broadcasting:* 5 February 1990, p. 37.
2. *Broadcasting:* 13 November 1990, p. 76.
3. Barry L. Sherman, *Telecommunications Management*, (New York: McGraw Hill, 1987), p. 186.
4. *Broadcasting*: 15 January 1990, p. 116.
5. *Broadcasting*: 20 November 1989, p. 90.
6. *Broadcasting*: 5 February 1990, p. 43.
7. *Broadcasting*: 20 November 1989, p. 44.

4

Media Economics

The organization of a media firm is affected by the type of ownership as well as its size, power, and market. Generally the larger the market, the more complex the tables of organization because there is more specialization and departmentalization in large markets. The more varied the type of ownership, such as group, conglomerate, or cable MSO, the more formal the chain of command and delineation of responsibilities will be.

Some stations may be small enough to occupy only one room, which houses the console, transmitter, a desk, typewriter, and file. These are all operated by one disc jockey, supported, when needed, by an engineer, who may be either the owner with engineering experience or an outside consulting engineer. In a true mom-and-pop station, the husband and wife share all station duties, including handling traffic, bookkeeping, and billing. On the other extreme, the large market station, especially television O & Os, may employ hundreds.

INSIDE THE STATION

Organizational Structure

Most organizations have upper management, middle management, and staff. A small market VHF or UHF may have fewer than 30 employees whereas a VHF network affiliate in a major market may employ as many as 400. It is important that a staff member know the exact responsibilities of a particular department head because job titles and responsibilities vary from station to station. Figures 2 and 3 show composites of a medium market radio station and a medium market television station. Keep in mind that these organizational charts are representative only; that each station may have a modification of these examples, according to top management's concept and experience. Stations that handle production will have larger staffs than those that are automated or network affiliated. Regardless of the detailed breakdowns, generally there are four fundamental departments at all media facilities: (1) general and administrative, (2) technical (engineering), (3) programming, and (4) sales. The same basic structure is found in noncommercial stations, as well, but the sales function is called development.

General and Administrative Services performed under this umbrella include accounting, legal, public relations, purchasing, personnel, and secretarial. The general manager is the person responsible to the owners and coordinates all departments and personnel in order to attain the station's objectives. Whereas personnel in other depart-

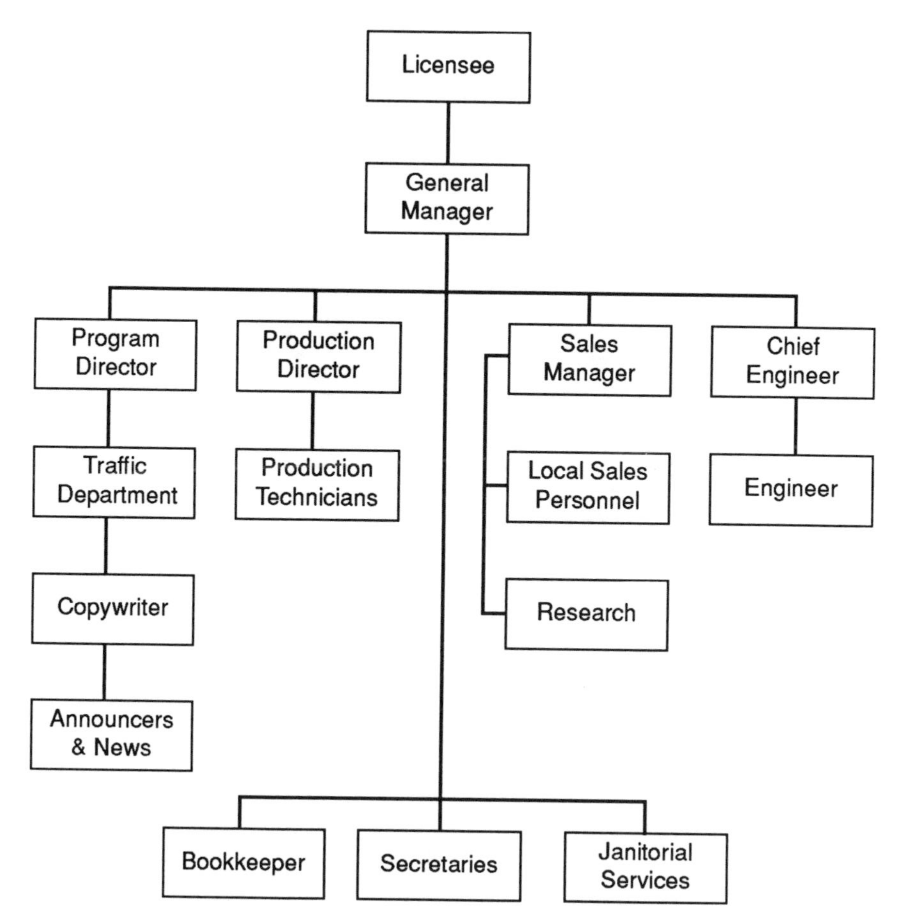

▶ ***Figure 2*** *Medium Market Radio.*

ments handle the nuts and bolts of operation, the general manager is responsible for planning, organizing, directing, and controlling. The general manager's highest priority is to establish the station's objectives pertaining to the station's financial position, the role of the station in programming as it is related to the community, and the performance of station personnel. The general manager seeks methods to keep expenses down and to achieve station objectives by raising revenue and reducing costs.

Programming Major programming policy and decisions result from consultations with the general manager, program manager, and sales manager. In addition to local live sources, prerecorded program material may come from independent program producers and syndication companies. The program manager is responsible for everything that is broadcast. On a daily basis, the manager supervises work assignments and schedules and handles budgetary programming matters and the problems of production. Program departments are responsible for staff talent, traffic, writing, announcing, art, and graphics.

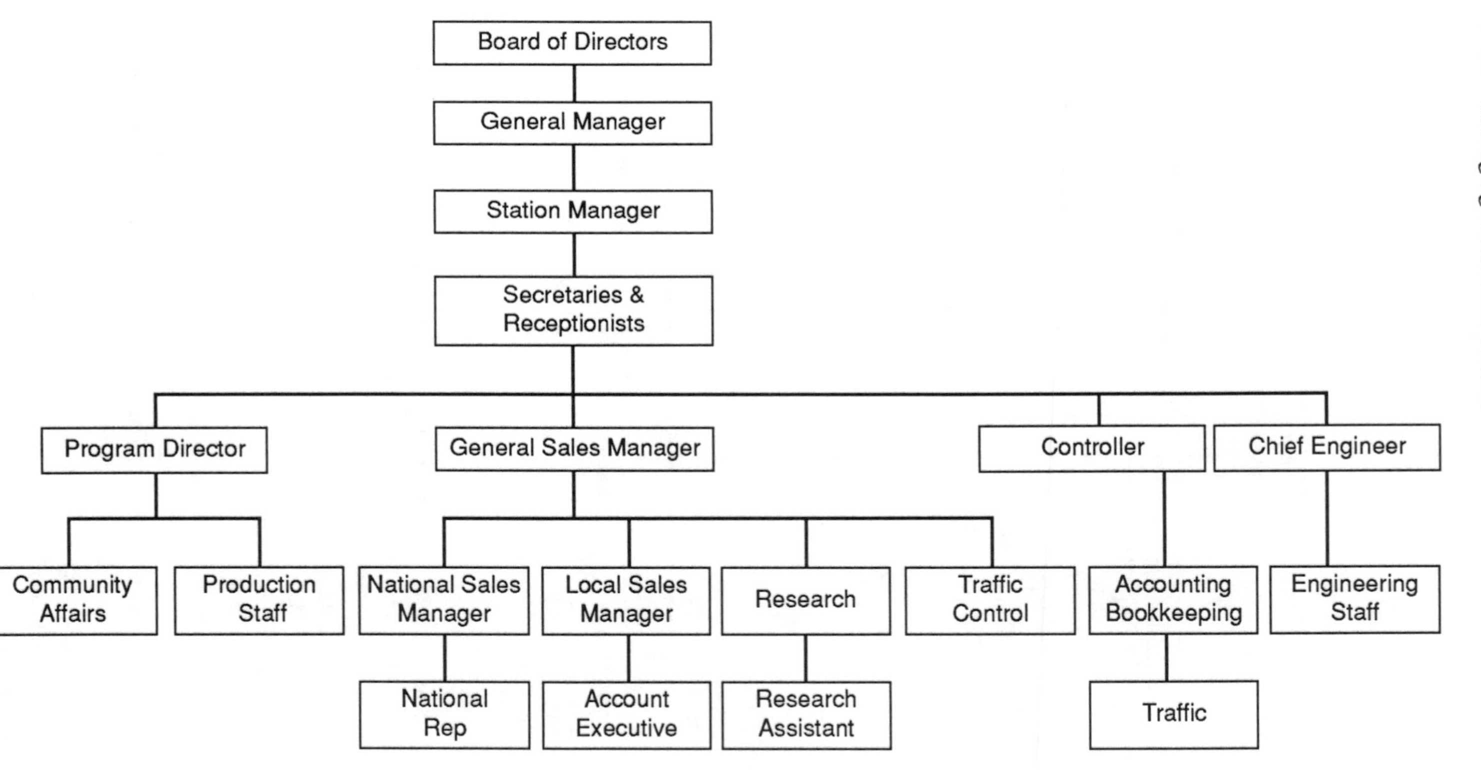

▶ ***Figure 3*** *Medium Market Television.*

News often falls under the responsibility of programming functions, but usually under a news director who reports to the program manager.

Sales The general sales manager normally combines selling with management. Sales is directly coordinated with programming since advertising must be spotted in or around programs that are most likely to draw the desired target audience. Programs as well as spot announcements are sold, and sometimes the sales manager helps create a program suitable for an account. As discussed in detail later in this chapter, the sales manager works closely with ratings and audience research methods and is the liaison with advertising agencies, station or national *reps,* as well as local clients (see Figure 4). In addition to helping the general manager establish the station's advertising policies and develop competitive rates, the sales manager also is responsible for developing supporting statistical and audiovisual sales tools for the sales staff to use in the field. Finally the sales manager must be familiar with retailing, merchandising, barter, and trade agreements.

Engineering Head of the engineering department is the chief engineer, who may supervise from 6 to 40 technicians. The position requires a fully qualified engineer who has spent a number of years as a working technician. Responsibilities include seeing that all equipment and transmitting functions are maintained at peak performance; and, in addition, planning and coordinating the engineering requirements of shows, including the assignment and scheduling of technicians. The chief engineer also makes recommendations regarding the purchase of new technical equipment. In addition, the chief engineer coordinates and negotiates use of outside telecommunications facilities with appropriate utility companies and private agencies; prepares all technical applications; and handles FCC paperwork relevant to authorization permits, license acquisition and renewal, or equipment modification.

Financial Administration

The *business department* handles a number of different tasks necessary to the functioning of the station. They include accounting, budget planning, billing, bookkeeping, and payroll.

Budget planning is an ongoing process and primarily deals with the future. Each year submissions of data are made by department heads listing anticipated revenues and expenses for the budget period. During the year regular budget reports are made to allow room for necessary modifications. Based on annual data the business department supervises cost controls and makes financial forecasting.

Writing an accurate budget must allow for unanticipated excess income as well as unexpected loss. To monitor the unexpected and to avert financial disaster, most stations use computers to provide periodic printouts on income and expenses, coded so that a running form of cost control is always available, enabling the accounting department to see immediately how each department is spending its share of the budget.

As part of the accounting function the business department compiles for top management a profit and loss (P & L) statement, usually compiled with breakdowns that reflect the costs of operating the four major departments discussed earlier in this chapter: general and administrative, sales, programming, and engineering. A final tally of expenses deducted from a tally of revenues results in the *bottom line* and reveals whether there was a profit or a loss.

Expenses While administrative expenses remain reasonably constant within the budgeting process, the cost of programming continues to rise. The only reason stations can afford rising program costs is that they are permitted to hold programs for a period of time and to repeat them. Programs owned by the station can be aired, after which the rights to show them can be sold to other stations or media.

Understandably, a television station spends about nine times as much on non-news programming as does a radio station because visual material is more costly than audio to produce. Likewise, television spends more on news than does radio. Whereas radio allocates approximately 5% of total expenses to news, a network VHF affiliate might budget 15% because television news requires a much larger staff, plus more expensive sophisticated electronic news-gathering equipment and satellite hookups.

In radio the major expense category is personnel, because the music product source is often free records provided by distributors in exchange for hoped-for exposure through airplay. Stations that subscribe to automated program services may pay more for the product, but usually need to employ fewer persons to handle programming.

In addition to the cost of running departments and meeting payrolls, additional expenses include cost of programs, payment for broadcast rights, music license fees, depreciation and amortization, employee fringe benefits, and station insurance, just as in any other business.

Professional external services also eat a portion of the income. Top management engages outside consultants, pays fees to professional associations, contracts for rating services, and engages communications attorneys as needed. Programming contracts run high for syndicated program or news services.

Revenue Sources A network earns most of its broadcasting income from nationally-aired commercials, called national spot sales. These commercials are simultaneously aired on network affiliated stations, which carry the network program. The affiliates receive from the networks a percentage of the income earned by network spot sales. Although O & Os must air programs fed by the network, affiliates are free to reject the programs if they wish. They only do so, however, if they can make more money airing a program on which they can sell more lucrative spots. Affiliates have an additional advantage in that the networks not only provide programs and arrange interconnections but also sell affiliate time in the national market.

Sometimes cash does not change hands at all. *Barter programming* is an arrangement in which an outside source provides a program, free of charge, but which contains a number of commercials advertising the provider's product. The program also contains open slots that the station can sell to local advertisers. In this system, the station is "paid" by receiving a program at no expense; in exchange the program provider gets free airtime for its advertising.

Another form of barter, called a *tradeout*, takes place when a station may exchange commercial airtime for merchandise or services. For example, a furniture store might provide office furniture in exchange for airtime of equal value.

Cable

Most of cable's revenue comes from subscriber fees. Many cable systems do little or no advertising sales, therefore, revenue from advertising is not significant.

ADVERTISING

The structure of broadcasting is strongly related to billings. Normally advertising accounts for about 75% of television revenues and approximately 25% of radio's income. To understand how advertising works, it is helpful to look at advertising agencies and how rates are established.

Advertisers

Advertisers are classified into three categories: (1) national, (2) regional, and (3) local. A *national* advertiser is one whose product, such as a popular soft drink or breakfast cereal, is distributed across the nation. A *regional* advertiser is one whose product may be available in some states, but not all. A *local* advertiser aims the sales message only to people within the immediate community. The nation's top ten television advertisers are identified annually in the trade publications, especially *Broadcast Advertiser Reports*. As a rule, among the top ten are Proctor and Gamble, General Motors, and Anheuser-Busch.

Advertisers generally do not place all their spots within a single show unless the program is a special that is known to attract a large audience, such as the "Super Bowl." To spread the coverage and the risk, today's advertisers use a scatter plan, buying time slots on carefully selected stations, which market research indicates are most likely to have the particular target audience most likely to buy the advertiser's product.

Some advertisers produce *cooperative advertising*, in which a local advertiser may pay only a portion of the cost of the spot, with the manufacturer of the product paying the remaining portion. Usually each pays 50%, but it can vary. Co-op advertising benefits all parties involved and is also a good incentive to encourage local merchants to buy airtime.

National and regional spot sales used to provide more income than did local, but in the 1980s a trend moved toward local sales. For 1990 industry experts anticipate a 4% to 6% revenue growth of spot and local combined with increased advertising expenditures from the airlines, fast food restaurants, and the movie industry.[1]

Advertising Agencies

Advertising agencies have been involved in broadcasting since 1923 when N.W. Ayer and Son handled advertising for the first sponsored show, "The Eveready Hour," on Station WEAF. At that time programs were prepared jointly by a station, the company, and the agency; the ads were mainly goodwill messages. Sometimes the name of the advertiser was incorporated into the title of the show, such as "The Lux Radio Theatre."

By the early 1930s, however, advertising agencies became heavily involved in producing not only commercials, but radio programs as well. In those early days of radio, it was the advertising agencies that took programming initiative, creating as well as producing the entire show; and they continued to do so until television arrived on the scene.

Agencies may be local, regional, or national. Sometimes an advertiser will establish its own agency, called a *house agency*, and will handle its own purchasing of time and do its own production in order to save paying the agency fee.

If an advertiser wants only a specific service instead of the entire range usually provided by a regular agency, it may buy only what it needs, such as research, public relations, creative ideas, or media, paying on a fee basis.

How a Buy Is Made

In a sense, today's agency is an extension of the client's own advertising department when planning and executing advertising campaigns. Figure 4 illustrates the interaction between the agency and the client and identifies the flow of the contacts.

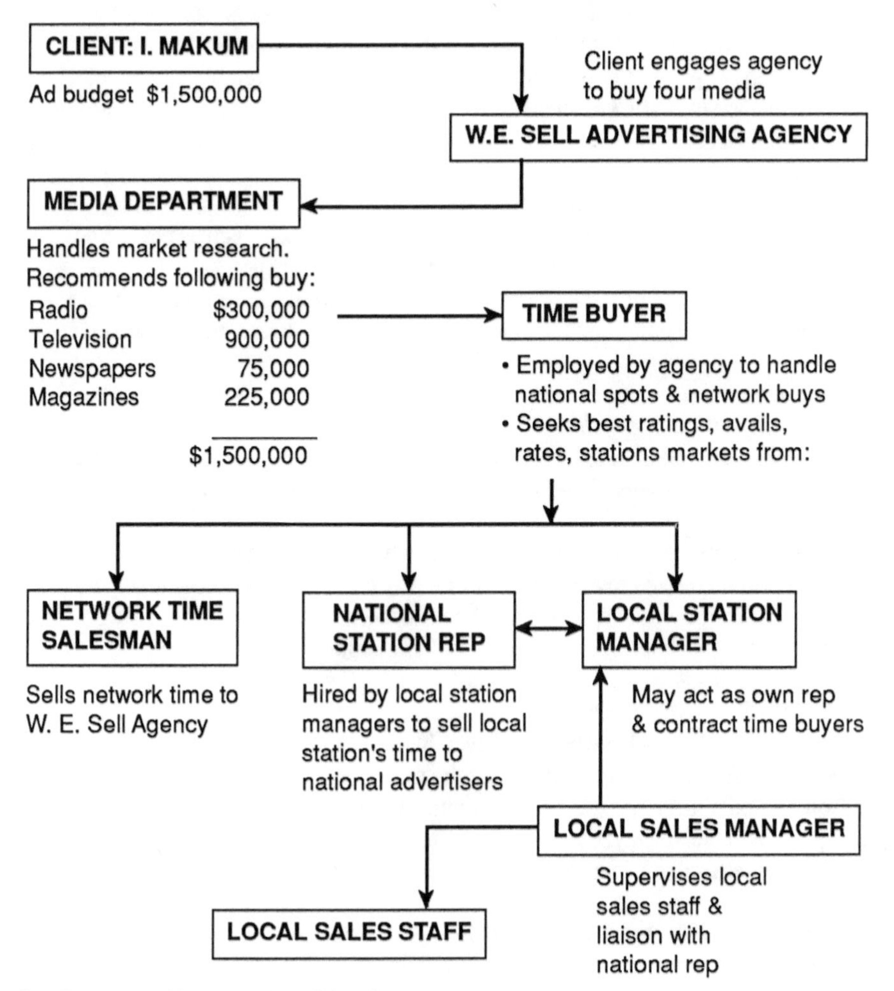

▶ *Figure 4 How to make a "Buy."*

Usually the agency charges the medium, not the advertiser, a fee or a percentage of the total billing. The standard 15% fee may be negotiated higher or lower. For illustration, let us assume that a hypothetical firm "I Makum" wants to introduce its new product "You Like" into a test city. It decides to use the W. E. Sell Advertising Agency to handle the account, and has budgeted $1,500,000 for the buy.

The W.E. Sell Advertising Agency assigns its own *media department* to handle marketing research, which comes up with the recommendation to put the money into four different media: newspapers, magazines, radio, and television, budgeting so that the total expenditures will match the final figure.

The media department then turns the finding over to one of W.E. Sell's *time buyers* who alone, or with buyers of the agency's print media, seek the best stations, their best availabilities, and the best markets; studies the programs, rates, ratings, and local market data; then buys (1) network time from the *network time salesman*, (2) regional time from a *station rep*, and (3) local time from the *station manager* (unless the station *sales manager* handles that liaison). Notice that in this instance, the *local time salesman* is not involved; that individual only sells time to local retailers and small chains within about a 50-mile radius.

When everything has been completed, the agency sends I. Makum a bill for $1,500,000. When the agency receives the money it keeps 15%, which in this case amounts to $225,000. The agency then sends the remaining $1,275,000 apportioned to the assorted media that ran the advertisements. The station and print media would have had to provide its own staff and time to perform similar services so, in a sense, the station has actually paid for the work done by the agency. And everybody benefits.

In this illustration we have seen that the *account executive* coordinates all departments and the marketing plan: the media department, research department, account group assigned to the particular client, and the creative department, which handles advertising strategy, copy theme, storyboards, and layouts.

Actual copy is often prepared by the agency's *copy department*. Art work for visual commercials on television or advertisements in newspapers and magazines is handled through the agency's *art department*.

A large agency has a separate *TV-Radio* department consisting of one or several people who plan campaigns, select programs for spot placement, write storyboards for commercials, may produce commercials and also buy time, determining when or where announcements are to be aired.

The *media department* recommends which media to buy, prepares media schedules and contracts, and buys appropriate space in print or on billboards.

At the station level, the sales manager keeps in close touch with the firm the general manager has contracted with to represent it in the national market. The representative company is usually called a *rep* for short. It is advantageous for a station to hire a station rep because the station itself cannot afford to have personnel in all the large cities that house advertising agencies handling most national and regional advertising.

The rep makes regular calls on agencies and advertisers with data about the local station, trying to encourage them to buy on the station he or she represents. Rep firms also offer advice to stations on matters to improve sales, programming, promotion, and research. The computer has begun to save some shoe leather and eased the updating of figures for such contacts.

Megareps Consolidation of rep firms called *megareps* began in the 1980s, and by the end of that decade two megareps, Katz Radio Group and Interep, were estimated to have control of some 90% of the national spot business. Station sales departments will have to adapt to the change.

Katz and Interep attained their leadership by developing complex sales structures and also by investing large amounts of money in research tools and research departments. In addition, they place heavy emphasis on the education and training of employees.

Rates

All broadcast stations and cable charge set rates for commercials and sponsored programs. In radio the rates vary according to the dayparts, with the highest rates during morning and evening drive times. Television also charges by its dayparts, which vary according to type of station, but with highest rates in prime time, and next to highest in parts adjacent to prime time. Target audiences are broken down into categories such as geographical location, population density, listening time, family size, income, employment, and gender.

Most sales consist of commercial announcements which may run 10, 15, 20, 30, or 60 seconds. They are called spot announcements because they are sold to run in specific spots on the schedule. A few spots may run as long as 120 seconds. In addition to selling spots, stations also sell blocks of time for entire programs.

Every station publishes a rate card. Discounts are often offered to clients in order to encourage buying of multiple spots on consecutive weeks. Local rates are usually lower than national rates.

Because there are usually fewer breaks in television programming than in radio, television sells time before, after, or within breaks in a program. These rates fluctuate according to the ratings and for this reason most television stations do not publish a rate card. Understandably, the more demand there is for a particular time slot, the higher the rate. Radio stations also sell time within certain programs such as sports, news, and daily features designed to carry spots.

Political Broadcasting Rates The FCC has rules and regulations concerning political broadcasting. Section 73.1940 of the Rules and Regulations affects how much a station may charge political candidates and limits the price a station may charge to be the "lowest unit charge of the station for the same class and amount of time for the same period." In addition, all candidates for political office are entitled to the same commercial discounts offered to commercial advertisers. The FCC also requires that stations keep available for public inspection all records of requests for time made by candidates and how candidates were given equal opportunities for airtime.

Cost per thousand (CPM) Advertisers pay as much as $100,000 for the average 30-second prime time network television spot and much more for a top-rated show or series. The 1989 "Super Bowl" spots, for example, cost $675,000 each, according to *Broadcasting/Cable Yearbook*, 1989. Rates on individual television stations may range from as little as $10 on a second-hundredth market and up to an average of $20,000 in

WPTF–AM ELECTRONIC RATE CARD ## A355MI26WSO

	MON	TUE	WED	THU	FRI	SAT	SUN
AMD → 6A-10A	$94	$92	$98	$116	$111	$99	$15
MID → 10A-3P	$71	$68	$76	$85	$91	$68	$15
PMD → 3P-7P	$60	$60	$62	$67	$67	$85	$15
NTE → 7P-12M	$21	$22	$23	$24	$25	$29	$15
ROS → 6A-12M	$65	$64	$74	$79	$81	$86	$42

DAYPART COMBINATION RATES

MON–FRI AMD	$102	SAT NUD + SAT PMD	$76
MON–FRI MID	$78	SUN MID + SUN PMD	$15
MON–FRI PMD	$63	MON–FRI AMD+SAT MID	$96
MON–FRI NTE	$23	MON–FRI MID+SAT MID	$76
MON–FRI AMD+PMD	$82	MON–FRI PMD+SAT MID	$64
MON–FRI AMD+MID+PMD	$81	MON–FRI AMD+PMD+SAT MID	$81

TOTAL WEEK Average rate $59

RATES SUBJECT TO CHANGE WITHOUT NOTICE — UNIT RATES

COPYRIGHT 1981 — MAXAGRID INC. — REVISED 1989

AUTHORIZED BY______________________________ DATE______________________

▶ ***Figure 5*** *Sample Rate Card. The rate card seen in Figure 5 illustrates the new approach to price spots based on supply and demand for each daypart. In this figure the copyrighted MAXAGRID, Inc.'s sophisticated computer program automatically updates WPTF-AM's rates based on availabilities. We see that WPTF offers single as well as daypart combination spot rates. This program utilizes the type of yield management used by airlines, hotels, and other services that are reserved in advance, and is based on early bookings to gauge the price-sensitivity of the market. WPTF-AM also utilizes a sales application from MAXAGRID, Inc., which enables salespeople to show advertisers how using different days/dayparts can actually increase the effectiveness of an advertising schedule on WPTF, based on reach and frequency. This program is now being used by over 200 broadcasters in the United States, Australia, New Zealand, Canada, United Kingdom, and Ireland.*

larger markets. Radio spots range from $1000 or more in major markets down to less than $1 in small towns.

We must keep in mind that costs cover not only the time slot of the particular station or network, but also the expense of producing the commercial, clearance rights for music, talent fees, and all aspects of station activity, especially those related to sales.

Because there is such a range of rates, a system was needed that would enable an advertiser to figure a statistically sound measure of what it costs to reach each thousand homes. This measurement is called *cost per thousand (CPM)* with the Roman numeral *M* meaning 1000. A favorable CPM is in the range of $3–$5, according to the market and the cost of the program. The goal is to calculate the number of people who were in the audience and exposed to that commercial. To illustrate, let us say that a station has an average audience of 800,000, and the rate for each airing of a commercial is $2000. The CPM would be figured as follows:

First, divide the audience figure by 1000 to determine how many thousands you have reached. Then divide that figure into the rate, as illustrated below. The result is the CPM.

$$\frac{\$2000}{800{,}000} = \frac{\$2000}{800\text{ (thousands)}} = \$2.50\text{ CPM}$$

In addition to using CPM to measure the cost to each 1000 households, it can also be used to measure the demographics of any particular group. CPM is also useful when comparing one medium to another; even one program to another. As a rule, advertisers and advertising agencies will not buy from a station whose CPM is exceptionally higher than competing stations. Whereas a good CPM used to be under $5 it can be higher as inflation raises costs.

When working with CPM it is necessary to know which medium is being referred to. For example, although it represents thousands of *persons* reached in radio, it can mean thousands of *households of persons* in television. If used to measure print it measures how many people bought the newspaper or magazine, but does not reveal how many read and were exposed to the advertisement in the magazine or newspaper.

One advantage the sale of spots has over retail sales is that once the cost of the commercial has been met, the remaining airings are profit. Thus, if a broadcaster airs a commercial for which he is being paid a rate of $1000 per airing, and if that commercial cost $20,000 to produce and sell, the first 20 airings will meet the cost with any further airings being profit.

MARKETS

Naturally, the larger the market the larger the potential audience. Advertisers want more than general "numbers." They also seek a particular demographic group such as by age, gender, buying habits, etc. Therefore an advertiser or agency seeking to make a buy will initially determine what kind of market it wishes to reach, by area as well as demographics. To help provide such data the two mayor U.S. rating services, Arbitron (ARB) and A.C. Nielsen (NSI) have apportioned the entire country into markets.

Arbitron identifies its 212 markets geographically, with no overlapping; then ranks them in descending order by the number of television homes they contain as well as the number of women, men, teenagers, and children in each, all identified by specific age spreads. ARB also provides station circulation data, explained later in this section.

Nielsen Marketing Research identifies and ranks television markets within each Nielsen Territory according to what it calls the Designated Market Area (DMA) and it reports sales data within the DMA. This offers national advertisers a way to determine the number of *spot markets* needed to achieve a given coverage of territory when buying time.

Market Definitions

Both ARB and NSI have devised market definitions that broadcasters use to identify the types of businesses to which they desire to sell time. A *home market* is that in which businesses draw customers only from the city of coverage; these are usually local retail or local services. If a firm's customers come from the metropolitan area surrounding the city, that market is identified as a *metro area*. In areas where the customers come from the suburbs as well as metro and home markets, it is referred to as a *designated market (DMA)* by Nielsen and called an *area of dominant influence (ADI)* by Arbitron. Should a business seek to sell to potential customers within the station's entire coverage area (even if it is more than one city), that is defined as the *total survey area (TSA)*. Maps indicating all ADIs and their rankings are published annually in *Broadcasting/Cable Yearbook*, which all commercial stations and sales departments use as a reference.

In addition, the yearbook publishes complete directories of every television station, radio station, and cable system in the United States and Canada, listing such information as each station's power, broadcast limitations, licensee, network affiliations, advertising rates, programming formats, and names of key personnel, as illustrated in Figure 6.

Ratings

Ratings usually determine which programs are kept on the air or removed. Although based on a sample, the percentage is projected to the total population represented by the sample. A rating is a *comparative estimate* of sets tuned to a program in any given market, or sets of markets. It is comparative in that a rating compares the *actual* estimated audience with the *total possible* audience. In a market with 10,000 households, a rating of 100 would indicate every home was watching. This, of course, is not likely to ever happen. If in that market 5000 people actually watched, it would be the equivalent of a rating of 50%. But it is not possible to measure all 10,000 homes; therefore, a sample of some 1200 homes (different services use different size samples) would be researched. If 600 people in that sample watched a particular show, that would be 50%; and thus projected to the total population of 10,000.

Another measurement, the *share,* is based on *households using television* (HUT). Instead of measuring the viewing of one station, the share estimates how many households in the particular market are watching a particular medium at the same time; in other words, a percentage of *all* television or radio sets in the market. In addition to

DIRECTORY OF RADIO STATIONS IN THE UNITED STATES AND CANADA

Key to the listings

Listed in this directory are all AM (amplitude modulation) and FM (frequency modulation) radio stations in the U.S., its territories and Canada. Information received as late as January 1, 1989 has been included in the listings. The following is a sample listing with explanation keyed to the boldface numbers.

> **WOF(AM)—(1)** Oct. 8, 1946 **(2)** 1,000 khz, 1 kw.D. 250 w.N. DAD (L-KQSL) (CP 5 kw-U) Stereo **(3)** Box 1000 (99999) (909) 555-1000 TWX, 909-999-9999 **(4)** General Broadcasting Corp (group owner, acq. 7-20-69) **(5)** Net ABC E. AP, Mountain State Network Rep. Jones & Company. Penn State Format MOR, C&W News staff 3, news progmg 14 hrs wkly Target audience: 35–64. Spec prog Sp 3 hrs weekly **(6)** John Jones, gen mgr; Mary Wells, gen sls mgr., Bob Frnch, prog dir, Alan Williams, news dir, David Smith, chief engr **(7)** Rates, $14, 13.50, 14, 12.50.
>
> **WOF-FM—(1)** October 1959. **(8)** 101 1 mhz; 3 kw Ant 300 ft. Stereo **(9)** Dups AM 50%. Format C&W **(10)** WOF-TV affil **(7)** Rates, $8.50; 7; 8.50; 7.

(1) Date station first went on air (regardless of subsequent ownership changes).

(2) Frequency in kilohertz; power and hours of operation. WOF operates with one kilowatt in daytime hours, 250 watts at night, and uses a directional antenna in daytime. Nighttime operation is limited to hours when dominant station KQSL is operating in daylight. WOF holds a construction permit for five kilowatts power day and night. WOF broadcasts in stereo.

(3) Address and zip code, telephone, Teletype Writer Exchange number.

(4) Licensee name and date of acquisition (if not original owner). If the licensee is a group owner—a company with several broadcast properties—it is so identified, as a group owner of which the licensee is a subsidiary. Details on group owners are listed in section A.

(5) Network, audio news service, rep and programming. WOF national affiliates are ABC Entertainment Network and AP Network. The regional affiliate is Mountain State. Sales representatives are also listed—national rep first, then regional. The WOF program format is part middle-of-the-road, part country and western, with three hours weekly of special programming in Spanish.

(6) Personnel.

▶ ***Figure 6*** *Sample Listing of Radio Stations and Explanation of Information. (Reprinted with permission from the Broadcasting–Cable Yearbook '89).*

(7) Advertising rates. The prices the station charges for a 12-times-a-week, one-minute spot are shown for each of for day-parts:
6 a.m.–10 a.m.; 10 a.m.–3 p.m.; 3 p.m.–7 p.m.; 7 p.m.–midnight in that order. If a day-part is applicable, a dash (—) replaces the rate.

Co-owned FM. WOF-FM has the same ownership as WOF(AM), and the FM listing contains only information different from the AM Co-owned AM and FM stations are always listed together, even when they have dissimilar call letters.

(8) Frequency is megahertz, effective radiated power and antenna height above average terrain. WOF-FM broadcasts in stereo.

(9) Programming. WOF-FM duplicates WOF programs 50% of the time and has a C&W format.

(10) Co-owned TV. WOF-TV has the same licensee as WOF-AM-FM.

Listings for independent FM stations follow the format of AM station entries. An asterisk(*) indicates noncommercial stations.

Operational abbreviations:

CH critical hours
CP construction permit
CR Canadian restricted
D day m1DAD directional antenna daytime only
DAN directional antenna nighttime only
DA-1 directional antenna day and night
DA-2 directional antenna with changing patterns day and night
DA-3 same as DA-2 with an additional pattern change
kw kilowatt
N night
SH specified hours of operation
ST shares time with
U unlimited time
w watt

share, radio also relies on *cume* figures (cumulative audience) which reveal how many *different* listemers were reached when all listeners are added together (cumlated).

Shares are always larger than ratings since they measure total audience of a medium. Usually share percentages are preferred as program measurements, while ratings are used as advertising sales tools. A show with a rating of 24.7 could have a corresponding share of 37.

Ratings measure the size of the audience in local and in national markets. The most commonly used measure is the ADI. Although ratings are only an estimate, they provide a useful statistical method of comparative analysis.

Coverage is the physical measure of the areas covered by the station's signal, and has nothing to do with the number of actual viewers and listeners. *Circulation* is a term which applies to the number of households (HH) in the country which tune in to a station at least once a week. Circulation only indicates a station's ability to generate a weekly audience, but offers no other data about the audience than numbers "reached."

Coverage and circulation data help a station to compute rates for both station and network time sales. They also provide comparative information for ad agency time buyers. It depends upon whether an advertiser wants to reach a large number of people, or a narrow market, as to which figures seem more important.

Radio data are cumulative and show the reach of a program over a set period of days or weeks.

More than 50 national commercial firms do audience research using four principal data-gathering methods: (1) telephone interviews, (2) in-person interviews, (3) listener/viewer diaries, and (4) meters. The services continue to try to improve methods to overcome deficiencies.

An indirect effect of ratings and sweeps is that when stations know in advance when they will be rated, they may air special programs or run contests and promotions which can inflate the figures.

Finally, good ratings do not always assure program retention. Stations often drop highly-rated shows because the type of viewers watching do not have the demographics desired by an advertiser. This is particularly true when a show receives high ratings from the over-50 group, because advertisers generally seek a younger target audience in its prime purchasing years.

On the other hand, quality shows are sometimes dropped because they do not have high ratings, even though they do have loyal, identifiable, listeners.

PUBLIC BROADCASTING

Public broadcasting stations are supported in a variety of ways. If licensed to an educational institution, a station receives some of its financial support from that institution; but such funds provide only the barest support. Additional funds are raised in creative ways such as on-the-air fund drives, auctions, selling ads in the station's printed program guides, and selling the station's production services. Another, and sometimes controversial, source of support is the practice of *underwriting* in which commercial sources provide some money in exchange for brief air credits. It is controversial in that some view the practice as just another form of airing commercials.

Many forms of support have been suggested, but the problem has been to insulate public broadcasting from political funding pressures. Congress agrees that what is needed is long-range funding, with stations needing to plan farther ahead than the current practice. The history of attempts to fund public broadcasting is complicated. At present the Corporation for Public Broadcasting (CPB), created in 1967 by Congress to oversee the funding and operations of noncommercial broadcasting, remains the federal overseer of noncommercial broadcasting. It supervises and allocates nearly half of the federal money allocated for public television through grants to program producers, stations, and regional networks. One major problem is that even though the Public Telecommunications Financing Act of 1978 established appropriations for annual distribution, Congress has seldom authorized the anticipated amounts. Under President Reagan, public broadcasting had another financial setback when Reagan said that the service should look more to the private sector and less to the federal government. At this writing public broadcasting is asking Congress to restore adequate funding, and Congress is expected to consider the topic before the end of the year.

CABLE

The cost to start a cable system can be very high. Laying cable costs from $10 a mile in rural areas to $100,000 a mile in urban areas: even as much as $300,000 a mile in localities where underground cable is required.

To equip a black and white operation may cost as low as $30,000 whereas a color studio setup may run to $200,000. Although 20% of all systems accept advertising on their local origination channels, most systems receive less than 5% of their gross revenues from advertising. Advertising rates on cable range from $2 to $300 for a 30-second spot.

Tiers

Most cable systems offer several levels of program service, charging subscribers a separate fee for each level. The first level, called *basic service*, includes all local television stations and perhaps one or more distance superstations or an advertiser-supported cable network.

The next tier may include pay cable channels. In this system subscribers pay a separate fee for each pay service chosen. Not all systems have multiple tiers, but there is usually a basic, an expanded basic, and a pay channel. Aside from contracting for packaged pay programs, like HBO, cable operators can lease a channel to a pay program operator or can obtain their own programming directly from the supplier, often using satellites for interconnection. By the end of the 1980s, pay cable was on approximately 7400 systems reaching 35 million subscribers in 50 states.[2] About 99% of cable's revenue comes from the monthly subscriber fees, with an average monthly fee of $14 for basic service.

Those cable systems which depend upon advertising for support require at least 25 million subscribers and should attract about 10% of that potential audience to make a profit. Few achieve those results. Successful systems include MTV, CBN, and Ted Turner's superstation WTBS.

C-SPAN is one cable network not dependent upon subscription fees or advertising for support. *Cable-Satellite Public Affairs Network (C-SPAN)* began in 1979 as a goodwill offering by the cable industry, and is supported by more than 2000 participating cable systems, which give up to 4 cents a month per subscriber to support it. C-SPAN operates 24 hours a day, carrying live coverage of governmental proceedings, including those of the House and the Senate, plus congressional committee hearings, call-in shows of public interest, and conferences. While not airing product advertising, it does permit corporate image messages in the same manner as does public broadcasting.

Cable Rating Systems

Efforts to rate cable and other new electronic media such as satellite networks, pay-TV services, and others were variable in the 1980s. Nielsen conducted special research as reported in its *Home Video Index* (HVI). Arbitron measured cable networks on individual market bases. They continue to seek appropriate methods because of the problem viewers have identifying on which of the many cable channels they watched a particular show. Entering the cable audience field in 1982, the Audit Bureau of Circulations (ABC), which is the official registrar of distribution counts for magazines and newspapers, also entered the cable audience measurement field to verify subscriber counts for advertiser-supported cable networks every half year.

In 1982 the Broadcast Rating Council changed its name to the *Electronic Media Rating Council (EMRC)* in order to reflect more accurately the growing forms of electronic media. The EMRC has established committees for television, radio, and cable, working informally with rating services as a liaison with the industry. It has developed a voluntary accreditation service for the electronic industry which means that if rating services submit findings to the EMRC for an audit and the results are found to be correct, the EMRC "double check" symbol may be used on the reported ratings.

SUMMARY

In this chapter we have examined the relationship between general organizational structure to the business of broadcasting and cable and their sources of revenue. Regardless of size, each station handles all of the following functions, although perhaps under different department heads. In general there are four areas of responsibilities: general and administrative; technical; programming; and sales.

The general manager is directly responsible to the owners, and is accountable for coordination of all departments and personnel so they can function efficiently, keep down expenses, and earn a profit for owners and shareholders. The program manager (sometimes called the program director) handles major programming policy and decisions, together with the general sales manager and the sales manager. The general sales manager wears several hats and is responsible for managing sales personnel, helping the general manager to establish advertising policies and rates, as well as serving as liaison between the station and advertising agencies. All technical matters fall under the eye of the chief engineer, who oversees all engineers, technicians, equipment selection, and maintenance. Financial matters, handled by the business department, cover budget planning, accounting, billing, bookkeeping, and payroll.

Broadcasting's income is based on the number of commercial slots sold and the rates charged for those spots. Most of a network's revenue comes from spot sales to national advertisers. While affiliates earn some revenue from those same spots, they also rely on regional and local advertising. Local stations depend primarily upon local advertisers, supplemented by regional and national advertisers.

Advertising agencies may be local, regional, or national. They provide a range of services for an advertiser: planning ad campaigns, selecting media, time slots, and support services. The interaction between the agency and station is detailed in this chapter.

Station rates vary according to the number of listeners/viewers in the coverage area, station ratings, and dayparts. The basic measurement is the CPM, which indicates what it costs the advertiser to reach 1000 homes.

Markets are identified by media research firms, predominantly Arbitron and Nielsen, which provide data that identify how many people watch what in which market, and then report their findings according to sales data within each designated market area (DMA). The ratings are an estimate of the number of sets tuned to a station in any given market, or sets of markets, comparing the estimated audience with the total possible audience. Other measures are share of audience and homes using television (HUT). More than 50 national commercial firms perform audience research.

Radio data are measured cumulatively and show the reach of a program over a set period of days or weeks, rather than of only one show as is measured on television.

Public broadcasting is funded partly by its licensed owner. Additional funds are also provided by Congress through the Corporation for Public Broadcasting (CPB), which supervises the federal fund allocations made through grants. However, the appropriated amounts have seldom been authorized, and public broadcasting is asking Congress to restore adequate long-range funding. Other funds are raised in creative ways, primarily through on-the-air fund drives.

To start a cable system can be extremely expensive. Revenue is earned through subscriber fees with a separate fee for each tier of programming the subscriber orders. The basic tier offers all local television stations; remaining tiers are forms of pay-TV. Cable systems are seeking methods to establish universal ratings, and some services are entering the cable audience measurement field.

Notes

1. *Broadcasting*: 12 February 1990, p. 19.
2. *Broadcasting/Cable Yearbook*: 1989, p. B-1.

5

Programming

The average American home watches television for 6 hours and 59 minutes a day, according to Nielsen Media Research statistics. There are more than 90 million U.S. homes with at least one television set, and of these homes it is estimated that about 63% of television homes are equipped with a VCR, while 53% are hooked into cable systems, according to Arbitron Television Research. In addition, there are approximately 527 million radio sets in the U.S., of which 65% are in homes while 35% are out of homes.[1]

The question is, how do stations compete for audiences? How do they make programming decisions? What competitive strategies do they use? In this chapter we will examine elements which go into programming decisions.

SCHEDULING STRATEGIES

Scheduling is a combination of science, art, and intuition. Although supported by the numbers of ratings, shares, cumes and demographic breakdowns of broadcasting and cable audiences, the challenge to the program director is to combine previous knowledge and experience with a sense of changing trends in audience preferences. We must remember that the purpose of a program is to attract the largest possible audience most likely to buy the products or services advertised within the program.

Effective scheduling controls and then maintains *audience flow*, particularly during breaks in programming. An audience which stays with a station is said to be maintaining *flowthrough*. If a member of the audience switches to a competing station, that action is a form of *outflow*. On the other hand, if people switch *to* a station, there is viewership *inflow*.

Audience flow may drop as much as 50% when a program distinctly different from the preceding one is aired. The program director's task is to determine whether there is justifiable risk in abrupt format change or whether the new program is likely to attract more, but different, people.

Basic program scheduling strategies include counterprogramming, hammocking, stunting, blocking, stripping, and cross-programming (also called bridging). As its name suggests, *counterprogramming* is used to attract an audience different from that of the competition's. For example, when NBC lost Major League baseball rights to CBS in 1989, NBC countered CBS with movies and miniseries based on books by Danielle Steel and Jackie Collins.

Cross-programming is a strategy used to keep audience flow by scheduling the strongest program to start on the air earlier than the competition does so that it captures the audience before the competition begins to air its prime-time programming.

To *hammock* is to place a new or weaker program between two strong ones in order to capture an audience for the new show. This technique was used to build an audience for the spinoff of "The Cosby Show" by placing "It's a Different World" in a hammock after Cosby's program and before "Cheers." Hammocking may also be used to recapture an audience for a show that has been slipping a bit in the ratings.

Stunting is the practice of interrupting regular programming with highly publicized special shows, a technique used most often in sweeps weeks rating periods.

Blocking originated with radio and was picked up also by television. Blocking is used most often by independent stations that must compete with networks and affiliates. The goal of blocking is to program a format, or a group of programs with similar appeal, to the same segments of audience. Radio continues to use it more than television does, airing blocks of music with the same format. Television blocks may consist of back-to-back situation comedy reruns, or blocks of movies, sports, or religious programming to counter the affiliates during non-prime-time scheduling.

Stripping is the technique of daily scheduling a local or syndicated series of programs to run at the same time, Monday through Friday. The stripping concept, borrowed from the comic strips, originated with radio's soap operas and has been picked up by television's soap operas. One advantage of stripping is that syndicated shows are usually discounted when purchased in large numbers and, consequently, the cost per show drops. Another advantage of stripping is that it develops audience loyalty, which may carry the listener into prime time viewing as well. It is also easier to promote a series rather than one single episode. The source of most strips are off-network series.

Cable uses stripping, blocking, and counterprogramming more than does television. Cable's superstations tend to counterprogram against the networks with sports and old movies. Repeat airing is an effective counterprogram strategy. Local cable systems frequently rerun scarce programs at different times of day and night, to reach different audiences. For the same reason, pay movie networks recycle movies on various days or weeks to reach folks with irregular schedules, and Cable News Network (CNN) repeats the same news hourly.

A practice not favored by the Federal Trade Commission and the FCC is *hyping*. Used during sweeps weeks, it is a practice of scheduling blockbuster programs instead of the regular schedule in an effort to boost the station's rating. Exceptional giveaway promotions during sweeps weeks is another type of *hype*. The problem is that these practices distort the ratings; therefore, the research services identify hype in their rating reports if hype could have inflated the rating figures.

DAYPARTS

The same nomenclature for dayparts is used by broadcasting, cable syndicators, and producers because of the need to have explicit references for buying time and setting rates.

Television's weekday dayparts revolve around *prime time*, which is 8:00 P.M.–11:00 P.M., Mondays through Fridays. The hour before prime time is called *access*, explained in Chapter 6 under the Prime Time Access Rule. *Late fringe* is from 11:00 P.M.–1:00 A.M. Following late fringe is *late night*, which is being viewed by some as the final frontier among dayparts. The period from 1:00 A.M.–10:00 A.M. is called *other*, followed by *day* from 10:00 A.M.–4:30 P.M. Following *day* is *early fringe* from 4:30 p.m. - 7:00 p.m. Sundays are slightly different, with *other* referring to all day up to prime time; and, with *late fringe* remaining the same as on weekdays.

Radio programming daypart segments are identified as *morning drive, midday, afternoon drive, night,* and *overnight.*

TELEVISION PROGRAM TYPES

The basic categories of programs include newscasts, documentaries, public affairs, quiz and game shows, sports, movies, situation comedies (*sitcoms*) variety (musical and comedy), general drama, action-adventure, soap operas, evening soaps, children's, and religious.

Entertainment

The Director's Guild chooses to identify creative programs in five categories: comedy series, dramatic series (night), dramatic series (day), musical variety, and dramatic specials.

Syndicated programs are identified as *off-network* versus *first-run*, referring to the program's source. A show which first was seen on the network and then put into syndication is called *off-network.* A program produced by independent producers, never shown on the network, is referred to as *first-run* when it appears on a non-network station.

Sitcoms are popular for syndication and daily stripping, but in 1990 even they showed signs of ratings slippage, perhaps reflecting a change in America's viewing habits or use of alternate technologies.

A *spin-off* is a new program which stars a secondary character from a hit show on the same network, such as "A Different World" spinning off from "The Cosby Show." Different from the spin-off is the *clone*, which imitates a successful program, but changes stars and minor plot backgrounds. "Dallas," for example, was cloned by many competitors.

Miniseries are special series which may run 8 to 12 hours long and are scheduled over two or three nights, generally a dramatization of a popular book.

Movies seen on television and cable may be either theatrical releases or made-for-TV movies, and are discussed in detail later in this chapter. Programs labeled *action-adventure* are usually stories involving police, detectives, Westerns, and war. Science fiction may also appear in this category.

Reality television gained strength in the 1990s with shows such as "Unsolved Mysteries," "America's Most Wanted," and "Rescue 911."

The financial underpinnings of prime-time television are the series, which consist of multiple episodes and share a common cast. Sometimes it takes a longer period of time for a quality series to generate good ratings, such as the hit program "Cheers," which took a long time to attract a sizable audience.

The current trend to abandon a program if it does not attain high ratings within several showings is a policy that is responsible for the small number of series available for syndicated stripping in the 1990 season.

News

A study by the Roper Organization, commissioned by the Television Information Office, shows that 66% of the U.S. public turns to television as the primary source of its news, with 55% ranking television as the most believable news source.[2]

Despite these findings, the prime-time news programs, outside of "60 Minutes" and "20/20" have lost viewers because of competition from CNN, cutting into what are called the *informational hours*. Viewers now have additional viewing options competing for their attention with cable, the emergence of fiber optic transmission, and the possibility of direct broadcast satellite service.

It may be that people spend viewing and listening time only on news events they consider most important. In January, 1990, news consumers spent less time on the major news stories involving changes in Eastern Europe. In contrast, the story that captured most attention was the crash of the Aviancia Airlines jetliner on Long Island, followed by the Charles Stuart murder case in Boston and the McMartin Day Care child abuse case in California. This, at a time when Dan Rather stressed the importance of foreign news coverage. Rather believes that the network news shows need to distinguish themselves from the competition, and suggests that at least one network could extend the evening newscast to 1 hour.

Obviously stations have difficult choices to make: the highly rated, cheap-to-produce reality shows, or the quality, expensive, prime-time newscasts.

Nonlocal news sources Television and radio affiliates receive news items and features over regular network relay facilities during slow parts of daytime hours when relays are not in use. The affiliates record the feeds and choose times for later insertion in the local newscast, paying the network a fee. Affiliates may also obtain the right to record regular network news programs as sources of stories for later use in their local programs.

Independent stations receive news for their local newscasts from national and international news providers such as UPITN and NIWS. *Independent News Network*, a syndicated news programming service, has initiated relaying packaged half-hour newscasts live via satellite to subscribing independent stations.

During the annual "Faculty-Industry Seminar," sponsored by the International Radio and Television Society for a group of distinguished journalism professors, it was predicted that the relationship between networks and local stations was *fracturing*, because a number of affiliates and O & Os had moved network evening newscasts out of prime-time access to program that particular daypart with local shows or with syndicated programs, since the non-news shows produce higher profits for the stations.

In addition, research indicates that viewers strongly identify with local news personalities and seem less interested, as a whole, in national or world events, preferring news about their own communities.

Much of what makes magazine shows appealing to network executives is economics. The production cost of "48 Hours," for example, is less than half of what it would cost to acquire a typical 1-hour network entertainment program.

Sports

Research indicates that the most popular professional sport on television is football, followed in popularity by baseball, basketball, tennis, and golf, in descending order. Attracting fewer viewers are auto racing, bowling, and soccer.

Because sports attracts a usually hard-to-reach demographic, middle-class males, they usually obtain higher than normal advertising rates.

For scheduling reasons, it is easier for cable and independent stations to air live sports on a regional and local basis than it is for broadcast networks and their affiliates. In addition, because of the higher costs to acquire broadcast rights, the affiliates usually receive no compensation for carrying the events. The local stations can, however, sell spots adjacent to the event, which carries national advertising. In the case of cable, some operators said they did not want any more NFL games from ESPN in 1990 because they had all the games they wanted, and that more games would cost subscribers more money at a time when operators are being criticized for rate increases. However, ESPN claimed it was the sports owners' raising the cost of the rights, not the cable network itself.

The "Super Bowl XXIV" had the lowest national ratings ever for a prime time Super Bowl, even though it was at the top of the ratings for its week with a 39.0 national rating a 63 share. However, CBS took in nearly $30 million in ad revenue, and gained a solid lead-in value for the premiere of the new series "Grand Slam," which earned a 16.6 rating and a 30 share. This illustrates the point that prestige may be more important than top ratings. Spenser Koch, general sales manager of WJBK-TV Detroit, identifies the "Super Bowl" as *event television* and believes that despite a drop in rating the show still reaches a substantial audience, and that the advertisers gain prestige for appearing in the "Super Bowl environment." He says, "It's an event, not a rating. You're not selling a spot in a show, but packaging an event."[3] Sports remains big money. In the first agreement of its kind, the United States Olympic Committee (USOC) signed a 3-year, $30 million broadcast and cable television rights agreement that by 1992 would result in 250 hours of Olympics events to be carried in the three major networks, as well as ESPN, WTBS-TV Atlanta, and SportsChannel America.

Studies of the effect of broadcasting upon sports reveal that in some areas the media are changing the game. The creation of commercial *time-outs* (breaks called about every 10 minutes by the referee to accommodate advertising spots within a game) is one example of how television has changed sports rules and strategies.

Another questionable practice affecting which games may or may not be carried is an agreement to delay the start of a game at the last minute, to provide time to sign up more stations or cable systems.

Finally, stations have been known to get around blackout rules, which specify that games must be sold out if they are to appear on television, by buying up the unsold tickets so the blackout rule will not apply, thus giving the station an opportunity to broadcast and sell advertising on the air.

Children's Programming

Although broadcasting's mandate to serve the public includes children's programming, there has long been controversy concerning what is "good" to counter the Saturday morning cartoons. Local stations used to schedule some type of children's programs on radio and television, during after-school hours or on Saturday mornings. But quality programs, especially on television, are expensive, time-consuming, and usually cannot equal the quality of network or studio production. There is, therefore, much less on the air than formerly. Cable has taken up some of the slack with children's channels. In addition Fox Broadcasting introduced the *Fox Children's Network.* The Discovery Channel also entered the interactive video market in 1990 by producing documentary videodiscs intended initially for teachers and educators and, eventually, for consumers. Supporting the discs are stills, graphics, and computer software, allowing the videodisc material to be controlled via computer. Disney also returned in 1990 with high quality animation in a show titled "Disney Afternoon."

An inquiry was in progress on Capitol Hill at this writing questioning why the government should not take a more active role to insure quality children's television programming. Senator J. James Exon addressed the National Telecommunications and Information Administration (NTIA) asking why broadcasting cannot provide competition for cable television systems that provide children's programming. Broadcasters feel that children's programming is generally a financial loss, and that they cannot commit heavily to such programming; yet, in the Neilsen's November 1989 sweeps, the results for syndication revealed that children's programs were doing particularly well.

According to child development experts, the negative impact of television on children's lives includes shortened attention spans in the classroom, insensitivity to violence, plus passive, nonparticipatory attitudes toward learning. In addition, children are conditioned by television advertising to ask for tie-in advertised merchandise which parents are pressured to buy. For these reasons, Action for Children's Television (ACT), a nonprofit group of parents, has petitioned the Congress and FCC for several decades to require more quality children's programming to be aired. However, in the deregulatory 1980s, a requirement for certain kinds of programming was considered as a violation of broadcasters' First Amendment rights to air what they wish, and consequently that provision was deleted. "Sesame Street," now in syndication, was one of the outstanding prototypes of this genre of programming. The heart of the controversy is whether economically disadvantaged children, who need programs of quality and education the most, would lost their source of free programs, because their families could not afford cable and pay-TV.

Made-for-TV Movies

In its early years television practically ruined the movie industry; many movie houses went bankrupt between 1946–1954 and closed down. Then things turned around as the major television broadcasters turned to Hollywood to provide them with films for programs. However, prime-time movies on television never really caught on until 1961 when NBC broadcast the movie *How to Marry a Millionaire*, featuring Marilyn Monroe. Success of the film encouraged other networks to follow NBC's move.

Networks, however, cut out portions they felt were inappropriate for home and family viewing, and filmmakers were upset at the cuts made to accommodate commercials. Generally, the movie studios at that time were not willing to sell their best work to television.

The networks discovered that buying the best popular films was financially extravagant. CBS paid $20 million for four showings of *Rocky II* even before the movie had proven itself at the box office. The price: $5 million for one showing!

From a declaration of independence by the networks was born the concept of made-for-television movies. Universal was the first and main supplier of these movies, pioneering the concept during the 1963–1964 season. In 1969 ABC introduced the "Movie of the Week," which consisted of original, 90-minute movies especially produced for television. At an average cost of only $375,000, it was much less expensive than a feature film. At first everybody thought the "Movie of the Week" would fail. It was the first show on television to discuss controversial topics, such as drugs, the Vietnam War, and even homosexuality. It was the start of movies that turned away from escapism.

Steven Spielberg is credited with legitimizing the made-for-TV movies by adding visual quality as well as good stories. Today we see made-for-TV movies not only as regular length movies, but also developed from novels as longer miniseries and docudrama. Ground breakers in the longer movie form were "Rich Man, Poor Man" in 1975, followed in 1977 by "Roots," which earned a high rating of 66.

Operation Prime Time produced a number of good made-for-TV movies, designed for independent stations, to be used as specials in prime time; and afterwards to be marketed to affiliates for stripping in access time. Pay-TV sought pay-TV rights for potentially strong theatrical films, and Showtime as well as HBO put up front money for films in script form. After first showing completed films on pay-TV, HBO, Showtime, and Operation Prime Time would then sell the films to the networks, and finally to stations.

Made-for-TV films have gained the same respect as theatrical films. In Europe, "Duel," "Tribes," and "Elvis" were distributed as feature films after being broadcast in the United States. Made-for-TV movies also help to launch new talent; for example, "Sybil" paved the way for Sally Field to make the theatrical film, *Norma Rae*.

Not all made-for-TV films are inexpensive to produce. The miniseries "Shogun," a 12-hour blockbuster shot in Japan, cost about $20 million. As a rule, however, such films are less costly than theatrical films for a number of reasons. They are completed more quickly with a shooting ratio of 2:1 compared to theatrical film's ratio of up to 15:1 if on location. Whereas theatrical films are not always adaptable for cutting to accommodate commercials, the made-for-TV movies are written in advance to avoid cuts that might interrupt the artistry or development of the story. Some *made-fors* serve as pilots for series which may come as spinoffs. And the producer of such movies can anticipate money that will be earned from foreign sales of the movies. In addition, stars usually ask for less money for such films because they consider the exposure valuable. Finally, a TV movie may be shot in several weeks, a far cry from the years required for some theatrical films. Thus, films made for TV are generally less expensive, compatible for broadcasting, and can be syndicated for additional earnings.

Competition from cable, however, could kill the golden goose. If pay cable continues to get exclusivity rights, and if the networks cannot afford exclusivity, there may be fewer movies shown on free television. In addition, ratings for syndicated repeat showings of movies on television have begun to drop because most of them have already been seen by too much of the television audience, making it difficult for movies to earn even enough revenue from advertising sold within the shows to pay the licensing fees. Television's main competition for the product is between the independents and the cable systems.

Minority Programming

As discussed in Chapter 3, black ownership of broadcast media started in the 1970s. By the 1980s there were several hundred black-owned radio and television stations, including the every-growing *Sheridan Broadcasting Network* (formerly the *Mutual Black Network*) covering the country, providing news and information to 114 commercial affiliated stations in approximately 130 U.S. marketing areas, with the potential to reach over 70% of the U.S. population. In addition SBN can be heard in 22 countries worldwide on 750 outlets via the Armed Forces Radio/Television Network. Since June, 1981, SBN has been airing a weekly feature entitled "Caucus Corner" from Washington with programs about matters affecting black Americans, such as voting rights. The cable Black Entertainment Channel also offers programs of cultural and popular music and arts. Community stations, whether minority-owned or not, now offer programs of interest to blacks, especially in areas where blacks are a significant portion of the community. Black ownership, however, does not imply strictly black programming. The owners say they program for the market they cover, and this includes programs which also serve the total area, because advertisers want the largest possible reach.

The second largest minority group in the United States, the Hispanics, are served nationally by the *Spanish Information Network* (SIN), which premiered June 1, 1981, and now airs its programs nationally over more than 100 satellite-interconnected outlets. Each week an estimated 12 million people watch Spanish language programs on SIN; and major advertisers, recognizing that the most effective way to reach their audience is in the language in which it thinks, gear specially written commercials to what may soon be the largest minority group in the United States. Rewriting of such commercials is sometimes necessary to avoid misunderstandings of meanings of trade names or of behavior considered unacceptable to the Spanish-speaking audience. The Spanish-surnamed people in the United States come from a number of different cultural groups: Chicanos, La Raza, Puerto Ricans, South Americans, and Cubans. With the growing Spanish-speaking market, management and sales aspirants should be fluent in the language and the culture.

Specialized Cable Stations

Cable has provided the means for an entire channel to be devoted to one type of programming. MTV and VH-1, for example, consist of short videos that dramatize popular music, particularly rock. These music channels not only compete with broadcast programming as a genre, but have also affected the way viewers aged 12–24 view television. One study indicated that those watching television programs will, during the commercials, "zap" to MTV and only return to the television program if MTV has not captured their interest, a practice which defeats the purpose of advertisers that buy time on television programs. The cable music channels have also cut into the audiences of rock radio stations. Thus cable music channels hurt both television and radio ratings.

Other specialized cable systems include ESPN, the all-sports channel; CNN's all-news channel; children's channels; and the two new basic comedy networks: HBO's ad-supported all-comedy channel and Viacom's comedy series "HA!"

RADIO FORMATS

There are 42 major radio formats being aired in the United States. Of the nation's growing total of almost 10,600 radio stations, approximately 8000 broadcast only one specific type of format, while more than 2000 report using two or more formats. About 500 stations offer three separate formats in different dayparts. A format is defined as programming broadcast more than 20 hours a week.

In 1990 the Birch ratings reported that the two strongest formats in the nation's two top markets (New York and Los Angeles) continued to remain Contemporary Hit Radio (CHR) and Urban Contemporary (UC). Another strong format was Album-Oriented Rock FM, as also were news and news/talk AM stations. Urban radio's top rating scored a 6.3 average quarter hour (AQH) share in 1989. Of the 2325 adult contemporary stations, about 50% are AM and 50% are FM.

Country and Western (C & W) formats are featured on more than 2000 stations, with about two-thirds on AM and one-third on FM. Noncommercial stations' formats include about 300 classical and 250 educational, followed closely by religious and jazz formats.

There is a general pattern of format preference that reflects the lifestyle and demographics of listeners. Contemporary is most popular in urban areas, contrasted to Country and Western in more rural sections, although such distinctions are not absolute. Stations choose format either because they had the opportunity to program the most popular sound, or because they thought it better to counterprogram in order to attract the audience from a different demographic or ethnic group.

Stations which air more than one format use block programming for different dayparts, scheduled to reach the audience segment most likely to buy the products advertised in each block. Generally it is the smaller stations which offer multiple formats. The most lucrative and most listened-to time periods are usually the morning and evening drive times. But drive times vary; therefore, there is no set national drive time period.

Narrower scheduling is found in competitive large metropolitan areas. In fact, some stations in the largest market have succeeded by giving airplay only to the top five or ten hit records of the week.

Decisions concerning what is to be aired may be either free form or rigid. If the program director or disc jockey has a pulse on the listenership, there may be great freedom of selection, perhaps scheduled according to blocks of sounds. Other stations have a very rigid format, repeated hourly, using a so-called *clock* formula, which indicates what type sound is to be aired at the same time each hour. Even with syndicated music, thanks to automation the music selections can be rearranged or spotted by the program director or disc jockey.

Format Syndicators

Format syndicators provide material for full 24-hour schedules for independent stations. This may include full schedules tailored to a particular sound, including appropriate features and fillers. In addition to on-air product, syndicators also offer advisory services as to merchandising and selling commercial time. Bill Drake, of Drake-Chenault Enterprises, is the oldest and largest format syndicator with more than eight formats geared to specific age groups. Other syndicators, such as Bonneville, offer "beautiful-music" service, especially for automated stereo FM stations and as background music for professional settings, such as the waiting rooms of physicians, dentists, etc.

More stations are turning to syndicated product, when affordable, because the normal former source of free records and albums, mailed by record companies to stations in hope of receiving air play, is drying up because of the increased cost of mailing and the cost of the records themselves. It seems to be most economically effective for a station to have inventory that is automated in some manner, with a system that gives easy access to a particular selection.

ECONOMICS OF PROGRAMMING

Costs of Programs

The cost to produce a television program is the same whether one thousand or one million people see it. Television economics has been a primary reason for the demise of live programming due to high production costs, which have forced the industry to use reruns and, consequently, the need to record them, usually on videotape. In fact, programming is often produced at a deficit in order to get an inventory of programs that can be syndicated later to national and international markets. The production method and salaries are the primary factors influencing the total cost of a program. Production includes direction, writing, talent, art, and graphics, as well as appropriate hardware and support operations.

Bookkeeping delineates between what are termed *above the line costs* and *below the line costs*. Above the line costs include all items related to the creative elements of production such as writing, directing, acting, and technical production. Below the line relates to physical or technical elements of a program such as production staff, scenery, costumes, equipment rental, editing, reprocessing, and overhead. About 45% of a program's budget consists of salaries for writers, performers, and directors. The remaining 55% goes for production items below the line. Additional costs include music license fees, copyright fees, and residual payments to performers, which are earned each time the show or commercial is seen on television or heard on radio.

Another expense is the cost of program distribution by a network. Network time charges are based on an *available* audience rather than the *actual* audiences. The average cost to interconnect 1 hour of time on a television network of 200 stations is about $200,000.

Syndication Economics

After a show is aired on the network, the showing rights are sold to a program supplier who then sells the showing rights to whomever they wish. These rights are referred to as a license.

About 95% of all prime-time television programming comes to the networks from outside suppliers, which may be the television departments of major studios, independent suppliers, or movie studios. We must remember that distributors do not usually engage in program creation, but they package the programs and handle the business arrangements. Even top hits seldom make a profit during the first showing. A key example is the hit series "The Mary Tyler Moore Show" which, although in top ratings for 8 years, barely broke even.

The program supplier may license other broadcasters or cable operators the right to air an agreed-upon package of programs for a fee. That license "sells" the right to show each episode a specific number of times over an agreed-upon period of time. Ideally, the price paid to show a syndicated program would equal the amount the station planned to charge for the commercials to be sold within the show.

It costs much less to obtain syndicated rights than to produce or buy a program outright. Not only are series syndicated, but also game shows, magazine shows, and talk shows, such as "Oprah Winfrey" and "Donahue." A station may be willing to pay for a top hit for two reasons: to air it, and also to prevent the competition from the opportunity to obtain it.

Barter Syndication By the end of the 1980s barter syndication had become a million-dollar business, and it was the barter element that drove syndication to new heights with gross national advertising totaling about $1.05 billion. Also increased were cash-plus-barter offerings. The result brought in a new source of revenue to finance production for the larger distributors. In fact, eight of the top ten shows in syndication had some sort of barter element attached to them. Barter programs are provided to the station free, but they already contain several commercials that the supplier has sold to national advertisers. The program also has several empty slots, which the station can sell for revenue it can keep. Sometimes there is a cash-and-barter arrangement whereby stations pay a fee for the programs as well as air some of the supplier's commercials.

International markets are a major source of income for producers of syndicated programs. Although the profit is seldom made in the first showing, the more often a show is rerun, the more money goes back to the producer. Without the international market for syndicated series reruns, few network prime time shows would be successful economically. "Star Trek," for example, was more financially successful as a syndicated series than it was as a network series.

Cable is a big customer for syndicated products, as are some satellite television program services. While cable penetration increases, viewing in television prime time decreases. In 1989 the total three-network share of home viewing was 67% with each network receiving no more than mid-20 shares. Syndicated products got a 28.7% share of the audience.

FIN-SYN Controversy The FCC's financial interest and syndication rules, (FIN-SYN), in controversy at this writing, may affect suppliers and networks. The rules prohibit the networks from acquiring financial interests in programming produced for them by others, and also forbid networks from competing in the domestic syndication market. If the rules should be repealed, they would allow the networks to produce for and participate in syndication; and would be competing with the independent producers for lucrative outside markets.

SYNDEX The FCC's new syndicated exclusivity rules (SYNDEX), which went into effect January 1, 1990, empower broadcasters to enforce exclusivity of programs against cable systems that import duplicate programming on distant broadcast signals, in order to protect the exclusivity that broadcasters bargain for in securing programming. Cable operators had challenged the ruling, but the U. S. Court of Appeals, in Washington, rejected their challenge. This has become interwoven with the *must carry rule*, according to which cable systems are required to carry local signals, since broadcasters have now introduced the possibility of charging cable for using their signals. The financial position of broadcasting is such that it needs protection from audience fragmentation, should a program they had bought and promoted and sold advertising for be seen on cable at the same time it was broadcast on the air.

Prime Time Access Rule (PTAR) The FCC's Prime Time Access Rule restricts the network's entertainment programs to a maximum of 3 out of any 4 of the prime-time hours (7:00 P.M.–11:00 P.M.). The purpose was to allow stations to use the *Access Time* from 7:00 P.M.–8:00 P.M. for community-oriented shows. But too often such shows had low ratings; therefore, stations began to fill the Access Time with locally produced or nationally syndicated *non*-network programs instead. The PTAR also forbids affiliates from airing off-network syndicated shows during the access period in the top 50 markets, thus freeing the time for independents and affiliates in the other 162 markets. Because the locally produced shows usually draw small audiences, the stations usually prefer syndicated game shows, which must be first run in the top 50 markets and must not be off-net syndication. PTAR in early prime time is less valuable for advertising, since the highest rates begin at 8:00 P.M.

Affiliate Economics Because affiliates receive about 65% of their programming from the networks, the empty slots to be filled are primarily early fringe and access periods. This limits affiliates' freedom to use most of the scheduling options discussed earlier in this chapter. Affiliates may reject a network program if they feel it is to their advantage to schedule something different in a particular slot, especially a program of high local interest such as a sporting event or major concert. Generally television affiliates spend about ten times as much on non-news programming as does a radio station.

News

Network Networks allocate considerable sums to news, as much as $300 million a year for hard news, public affairs, documentaries, political coverage, and unscheduled events, such as disaster coverage. In addition, salaries for staff and management, if not related to any particular program, may total up to $35 million annually.

It costs television much more to produce a news program than it costs radio because of the necessary additional personnel and equipment. A network-affiliated television station allocates about 15% of its budget to news as compared to radio's allocation of 5%.

News has been an extremely important product for television, ever since people began to turn to this medium as their main source of news in the 1960s. Increased viewing attracted advertisers to news programs. The question at this writing is whether the competing technologies will take away the economic advantage of television news.

Local Local television program production today is almost exclusively limited to news. Locally produced entertainment programs usually make much poorer showings in competition with syndicated or network shows. Stations in larger markets do more local production than do smaller stations, especially in access time. Most radio programming expenses go for personnel.

Cable

It is estimated that some 4400 systems originate programming in their own studios. Additional sources of programs are packaged pay programs (like HBO) or cable operators who can lease a channel to a pay program operator. They may also secure their own programming directly from a supplier. Cultural cable channels compete with noncommercial PBS for rights to many of the same performance events such as dramas, ballet, dance forms, music, opera, and literary works. Cable also airs foreign films, especially classic French, German, and Italian art films, in addition to recent foreign-made movies. David Kenin, senior vice president of programming for USA network, anticipates that in the 1990s cable will be viewed as a necessity in the same way that television is today. Cable networks will continue as a major thrust for cable. In Kenin's opinion, broadcast networks will continue to adapt, and there will be more cooperation between broadcast and cable networks, as seen by the Olympics deal put together for 1992 and 1994.[4] Economic realities are going to force more cooperation among production and distribution entities.

AUDIENCE ANALYSIS

Methods

There are many non-rating methods of audience research besides those that reveal numbers of viewers and their demographics, as discussed in Chapter 4. Psychographic research focuses on the needs, interests, attitudes, and general lifestyles of the media audience, sometimes correlated with levels of education and ethnic culture. The industry likes to find out what people like or dislike, what they recognize and remember, what they overlook and forget. Attitudinal studies seek to discover people's reasons for viewing.

Methods using a *program analyzer* research programs and commercials by equipping each member of the research audience's seat arm with buttons that can be pushed to indicate what is liked or disliked during the showing. The TV-Q evaluates the familiarity and likability of a performer, considered very important not only to radio, but also to television news. The latter's value is sometimes controversial because some believe that too much emphasis is placed on a newscaster's delivery, hairstyle, and the design of the set. Personality studies are extremely important to radio programmers who are interested in assessing how an audience perceives not only the station personalities but also the music playlists.

Although voluntary mail used to be considered an important type of audience judgment, research reveals that letter writers differ from the general population in terms of education, race, income, job, age, and marital status. Also, mail is sometimes obviously the result of one pressure group. Stations tend to pay most attention to original, strong letters of complaint.

Non-rating forms of research are important tools to supplement numbers. While sales personnel are most interested in estimated audience size indicated by rating, share, cume, and AQH, programmers are more interested in what attracts or loses an audience. Three forms of qualitative audience analysis are focus groups, call-outs, and auditorium tests. *Call-outs* is a term used for a telephone survey targeted to samples of station listeners or non-listeners, identified as letter writers or contest participants. This is used primarily for radio.

Research Service

Resources available to radio, television, and cable for help in developing sales are available from the Radio Advertising Bureau (RAB), Television Advertising Bureau (TVB), and Cable Advertising Bureau (CAB), all of which gather demographic, psychographic, and consumer purchasing information of value to industry organizations involved in advertising, promotion, and public relations. The major networks also have research departments as do the major cable networks and MSOs. The NAB has issued a series of reports useful to managers and sales. Names of commercial research firms and their services are found in the major trade publications.

The 1980s saw the development of many new technologies. The challenge of the 1990s is to create new programming and to do it better in order to catch up with technology that is outstripping programming. Programmers may have to take more risks and break into new programming concepts and styles. The most serious problem is lack of finances to support innovative programming. Broadcasting, once considered a "license to print money," does not seem to be able to make enough money for shareholders any more. Continued economic inroads are being made by fiber optics and HDTV. It appears that the 1990s may be a turning point for electronic media's relationship to programming.

SUMMARY

This chapter focuses on program practices and concerns that programmers face every day. Obtaining or creating a good program is only part of the battle. The key to success or failure is the strategy of program placement in the most likely daypart and broadcast time. Effective scheduling is that which controls and maintains audience flow using a variety of strategies such as counterprogramming, cross-programming, hammocking, stunting, blocking, and hyping. Cable uses blocking and counterprogramming more than does television.

Broadcasting, cable, syndicators, and producers all use the same nomenclature for dayparts when buying time and setting rates.

Programs may be classified either according to their content—such as news, quiz, sports, comedy, drama, action-adventure, children's, and religious— or as to whether they are first-run or off-network. Syndicated shows are generally stripped and may be the only way a producer makes a profit. Movies seen on television and cable may be theatrical releases or made-for-TV. A new format, Music Television, which has cut into both television and radio ratings, appeared on cable in the 1980's.

Series are the financial base of prime-time television and the successful ones go into syndication for stripping. Sitcoms were the most popular series in the 1980s, followed closely by action-adventure shows.

There are 42 major radio formats on the air in the United States as well as a number of specialized and minority formats. The top sounds in the major markets are Contemporary Hit Radio (CHR) and Urban Contemporary with Country and Western (C & W) in third ranking. Some radio stations air more than one format, choosing to broadcast the different sounds in selected blocks of time. Many radio stations are turning to format syndicators and automation, freeing the disc jockey to do production and creative work in assigned periods of time.

Because of the high cost of live programming, there have been fewer live shows and more reruns. The method of production and salaries are the primary factors influencing program costs.

Barter syndication was on the rise in 1990 and was behind the rising financial success of syndication by bringing in new national advertising. International markets are also a major source of income for producers of syndicated programs. Cable is also a big user of syndicated programs. For these reasons, a producer is willing to air a program for the first time at a loss in order to gain an audience for a show, which later can be syndicated, stripped, and sold to many outlets. Legal matters which may affect syndication are FIN-SIN and SYNDEX.

There are non-rating methods of audience research of a qualitative aspects to enhance the rating methods. The biggest challenge to broadcasters of the 1990s is to develop new and competitive programming to serve itself and the new technologies.

Notes

1. *Broadcasting/Cable Yearbook*: 1989, p. A-3.
2. *Broadcasting/Cable Yearbook*: 1989, p. A-3.
3. *Broadcasting*: 5 February 1990, p. 29.
4. *Broadcasting*: 1 January 1990, p. 37.

6 Regulation and Controls

Although direct government censorship of broadcasting is forbidden, there are many regulatory bodies and social forms of control which interact with broadcasting and with which management must be knowledgeable. Top management, of course, engages communications lawyers to assist with updating on regulatory matters, but there are also public and ethical aspects, in addition to First Amendment rights, that often provide more questions than answers.

GOVERNMENT REGULATION OF BROADCASTING

Antitrust Laws

As discussed in Chapter 2, broadcasting is a form of "commerce" in the sense that A T & T originally wanted broadcasting defined as a common carrier; subsequently broadcasting became responsible to the Secretary of Commerce. It is under the concept of commerce, therefore, that antitrust laws apply to broadcasting.

The main purpose of the two antitrust acts, the Sherman Act and the Clayton Act, is to prevent creation of monopolies. Courts view the antitrust laws as ways to promote the First Amendment's protection of competitive ideas. With new emphasis placed upon market-based deregulation, and with the rise of conglomerate ownership, there may be increased focus by government upon potential antitrust relationships in the 1990s, especially in the matter of vertical integration, which can permit one group to control all phases of production, thus limiting the concept of diversification of ideas and opinions.

It is for this reason that television networks are forbidden to own syndication rights to programs which they broadcast, and why television stations may not own cable systems. For example, major cable channels, including HBO, Turner Broadcasting, and Showtime, have branched from distribution into original productions. Moreover, television, advertising agencies, and rep firms have joined to produce original programming under such ventures as *Operation Prime Time*. This situation is being studied by Congress and may affect production and broadcasting in the 1990s. Vertical integration is good if it helps regulate costs and streamline expenses. It is bad if small groups of companies control all phases of production, thereby controlling information.

Horizontal integration of business differs from vertical in that the firm spreads its influence horizontally across a range of geographic markets, such as in the television/

affiliate relationship. It is also the form found in cable MSO franchises. When there is horizontal integration, firms may diversify and expand into related and semi-related enterprises, which is why networks are permitted to own magazines, record companies, and even amusement parks and food companies.

Regulatory agencies such as the FCC and the Securities and Exchange Commission keep a watchful eye on acquisitions and mergers, which might have so few sellers that the actions of any one of them could materially affect price; and thus have a measurable impact upon competitors, so that profits above normal may persist.

Advertising Regulations

The Federal Trade Commission Act of 1914 was originally intended to protect businesses from unfair competition. However, in 1938 the FTC Act was amended to protect consumers from unfair business practices. This established the FTC's basis to police deceptive broadcast advertising, even though no harm to a competitor could be proved. The FCC has no authority over broadcast advertising; however, it expects broadcasters and cable operators to cooperate with the FTC by withdrawing any advertising the FTC may find in violation.

Obscenity and Indecency

The intention of the First Amendment is to maintain a marketplace for ideas. However, speech which is obscene, defamatory, invades privacy, or incites insurrection is *not* protected by the First Amendment, because such words contribute nothing to the marketplace of ideas and, as such, is *unprotected speech.*

Under the First Amendment people are protected from censorship (prior restraint) by the government; however, the First Amendment does not protect people from people. Therefore the FCC, as part of government, has no power of censorship over the communications transmitted for the general public. On the other hand, it is not censorship if program decision-makers choose to edit material before it is aired, since this is a form of people making the decisions, not the government.

Based on the 1973 *Miller* case, current obscenity law is limited to hard-core pornography, since community standards are difficult to determine, varying from place to place. Generally the law requires the courts to consider the taste of the average person, applying contemporary community standards to determine whether material is obscene. In broadcasting, the FCC defines *indecency* as material that does not meet community standards for *broadcast media.* Based on this interpretation, the 1989 Congress stated that it thought the FCC should make it a top priority to regulate indecent as well as violent programming. In response, the FCC unanimously adopted a policy that prohibits "indecent broadcasts 24 hours per day." That policy and its underlying laws were called into the courts on First Amendment grounds by a coalition of broadcasting and public-interest groups and remains unresolved at present.

Concerns about indecency and violence arise because broadcasting is different from print; there is the possibility that children may be watching or listening to adult-oriented material on the airwaves. For this reason broadcasters are urged to be sensitive to the time of day such material is scheduled. What makes this more difficult is that research shows that children watch not only during the formerly "protected" after-school hours, but also in prime time. Access to cable channels increases the possibility of a child seeing material which might be deemed indecent.

Cable is regulated by states as well as by federal government. However, cable is different in that it does not use the public's portion of the airwaves. Nevertheless, prior restraint by regulating bodies would be considered censorship. Therefore, as a form of punishment "after the fact," the Cable Act of 1984 provides for fines up to $10,000 or two years imprisonment for anyone who "transmits over any cable system any matter which is obscene or otherwise unprotected by the constitution."

Because Congress has again raised the question of indecency on the airwaves it remains to be seen how matters related to obscenity and indecency on radio, television, and cable may be handled in the future.

Fairness Doctrine

Another issue unresolved at this time is the future of the Fairness Doctrine. A brief background of the doctrine may help put the matter into perspective. Although repealed in 1987, the 30-year-old statutory amendment may reappear in the form of a bill to be passed in 1990 by Congress. The issues, themselves, should be of interest to future broadcast management. Through the *Red Lion Broadcasting v. FCC* (1969) the Supreme Court unanimously upheld the constitutionality of the Fairness Doctrine as protecting First Amendment rights, under the principle that because the airwaves belong to the people, citizens should not have to beg for their rights.

Under the Fairness Doctrine, broadcasts were required to cover controversial issues of public importance in a balanced manner, and to assure reasonable opportunity for all responsible voices in a community to be heard. To seek to offer diverse views and facts, the doctrine covered fairness in areas of personal attack, political editorializing by a station, and controversial issues in general. It meant also that if a person were attacked on the air, that person had a right to respond. Concerning political broadcasting, if one candidate for an office were given free time, then all other candidates for the same office were entitled to the same amount of free time.

The FCC supported the Fairness Doctrine with the argument that regulation in broadcasting is needed since a station cannot expand its 24-hour day the way a newspaper can add pages. It was perceived from the viewpoint that although stations belong to private individuals, the airwaves they use belong to the public, whereas newspapers are private and do not use the public's facilities. Broadcasters felt constrained under the Fairness Doctrine since they found it was often difficult to provide for all viewpoints, consequently they generally tried to avoid discussion of controversial subjects except in *bona fide* newscasts and interviews. The Fairness Doctrine was considered legally acceptable since the First Amendment does not forbid regulation; it only forbids abridging freedom of speech.

Fairness also applied to controversial advertisements. In general the unanswered questions were: Fair to whom? To the industry? To the public? To those who seek to air their views? In a surprise move in 1987, the FCC repealed the doctrine after much study, reporting that it had "grave legal doubt on the Fairness Doctrine" and believed that the doctrine "violates the First Amendment and contravenes the public interest."

Media Access Project, a public interest law firm, challenged the FCC's reversal in 1989. At this writing it is anticipated that Congress and additional citizen groups will consider writing a fairness bill of some type into law in 1990. The Media Access Project and others support the return of the doctrine, claiming that the commission's decision to repeal the doctrine "subverted the public trustee model of broadcast regulation and substituted a print model in its place."

Copyright Laws

Copyright regulation is based on the 1976 Copyright Act, according to which payments are apportioned to holders of copyrights, but also allows limited *fair use* of copyrighted material with no payment. *Use* is legally defined as making something public through performing, displaying, publishing, etc. Copyrightable works include books, musical compositions, motion pictures, broadcast programs, art works, computer programs, and choreographic dance notations. A copyright is good for the life of its creator plus 50 more years, after which the work enters public domain and can be used without having to obtain permission or to pay royalties.

Broadcasting and cable are involved with copyright payment in matters related to airing music, dramas, and programs produced by others. In the 1980s cable was required to pay a compulsory copyright license for local and distant signals, but the FCC considered abolishing it in the 1990s, saying that the move would benefit consumers, broadcasters, and cable program services by reducing prices. Cable is automatically given retransmission rights for material lawfully retransmitted from television stations licensed to use the material, and pays a license fee for such use.

Performance Licensing Creative people and production departments must be knowledgeable about current copyright law, particularly those at smaller stations which may be producing their own spots or programs. They must know what their station is licensed to use and to be able to identify what is genuinely in public domain.

Movie theaters are required to buy a public performance license, as also do institutions that show home videos to groups of people—even libraries, colleges, churches, and prisons. The principle is that if a movie is shown in individual rooms, it is not a violation of the copyright law; but if it's shown in a public accommodation area where large numbers of people can see it together, copyright protection applies.

The three music licensing organizations are ASCAP, BMI and SESAC. Each of these licenses individual song titles. To ease clearances, most stations hold a blanket license from one or more of these three rather than pay per use, and those holding blanket licenses may use any music in their catalogs in return for payment of an annual percentage of each station's gross income. Not all stations use blanket licenses. Those radio stations, for example, with predominantly talk formats, pay on a per-use basis since they do not use enough music to justify the blanket fee.

Unethical Practices

The sponsor identification law makes practices such as payola and plugola illegal; the responsible person is subject to a fine of $10,000 a year or one year in jail for each violation.

Plugola occurs when an employee promotes on the air some product or event which would be of financial advantage to that individual. One celebrated case concerned a disc jockey who also had money invested in a well-known music group. Just prior to an upcoming live concert with paid admissions, the disc jockey broadcast recordings by that group as a subtle form of free plugging for the concert.

Payola is a practice in which a recording company representative makes secret payoffs in the form of money, drugs, big trips, or expensive gifts such as cars or fur coats to those disc jockeys or even program directors who make decisions about which selections to air, encouraging them to give extra airplay to that company's releases. Top management is responsible for establishing monitoring systems to prevent such illegal activities.

Federal Communications Commission (FCC) Regulation

Distribution Systems As discussed in Chapter 2, the most pervasive type of regulation is obviously the FCC's control over spectrum and technology, the primary purpose being to prevent signal interference, overlapping, and to provide the best quality in all types of communication distribution systems used by the public, from telephones to satellites.

The FCC is thus responsible for determining which standard or system to approve when more than one is under study, such as the current examination as to how best to provide HDTV service in view of limited spectrum availability.

Multichannel Television Sound (MTS) When possible, the FCC sets a standard such as it did when adopting MTS; and then allows broadcasters to use any system on the market, provided that the system does not interfere with the standard. This approach prevents FCC interference with broadcasters' equipment choice, yet simultaneously encourages marketplace technological improvements.

Multipoint Distribution Service (MDS) Multipoint distribution service, which began in the early 1970s, proved an effective way to distribute pay programming to apartment complexes, hotels, and similar multiunit dwellings. As the service developed the FCC amended its rules and added more channels. When an MDS system could offer more than ten channels it began to identify itself as *wireless cable* because it gave customers access not only to MDS programs, but also to ITFS signals, improved UHF reception, and local VHF stations; thus, it began to compete with cable television in a small way. The FCC authorities monitor and authorize each system separately to assure that satisfactory technological standards are being met.

Equal Employment Opportunities With the passage of the Civil Rights Act of 1964, which prohibits discrimination in employment practices by any firm with 15 or more employees, the FCC followed in 1969 by incorporating federal Equal Employment Opportunities (EEO) requirements into its own rules, applicable to both broadcasting and cable. Reports on employment are made to the FCC at license renewal time. The philosophy behind EEO in broadcasting is that in a democratic society made up of diverse cultural groups, each group's voice and ideas should have the opportunity to be shared.

CABLE REGULATION

Regulation of cable was in a flux at the start of the 1990s. Back in the late 1970s and early 1980s, when municipalities exercised wide-ranging control over cable rates and services, regulation of cable was felt by operators to be too heavy-handed. Since deregulation in the mid-1980s, regulation of cable was seen as too permissive.

The FCC started to regulate cable television in 1965 by adopting the same rules that existed for CATV systems served by microwaves, requiring operators to carry the signals of local stations and to refrain from duplicating within 15 days the programs broadcast by local stations. The following year the FCC initiated the "must carry" rules, and cancelled the 15-day protection. It was not until passage of the Cable Communications Act of 1984 that Congress provided some formal guidance and limited federal and local controls on cable operators.

Prior to that deregulation, the basic cable tier of service consisted largely of broadcast signals. But after passage of the Cable Act of 1984 and the FCC's deregulation of basic cable rates, cable operators began to shift cable programming services such as superstations and ESPN to the basic tier to make it more attractive to consumers, because it was felt that local broadcast signals were no longer a good substitute for the full range of programming then included on the basic tier.

In addition, to minimize copyright liability for carrying distant broadcast signals, and in anticipation of rate regulation, in 1989 cable operators had begun to strip basic tiers of all but broadcasting signals, placing cable programming services on the expanded basic tiers beyond the reach of regulators, since it was anticipated that only basic would be regulated.

By 1990 pressure rose from Congress and from broadcasters to reregulate cable, particularly to reinstitute the matter of the "must carry" law. Congress was also preparing to discuss whether cable operators should pay the broadcast industry for carriage of its signals. The third major issue under study by the Communications Subcommittee was the matter of vertical integration within the cable industry due to concern about concentration of ownership within the media.

The main justification for regulation of cable lies in the very nature of the industry. Most cable companies do not operate in classic free markets. Instead, they enjoy a type of monopoly in the communities they service, sometimes with exclusive franchises.

FCC chairman, Alfred Sikes, sees cable systems as unregulated monopolies, and his goal is to encourage competition with cable so that the government does not have to "substitute regulatory action for the market."[1] Sikes feels that action is needed. He sees the 1984 Cable Act as limiting potential competition between cable and other video transmission facilities. The FCC's goal, according to Sikes, is to foster multichannel competition and to create a video marketplace in which cable systems would compete with other multichannel video providers, including telephone companies, home satellites, program packagers, and wireless cable. Meanwhile, he believes that rate regulation may be needed during the transition to the competitive market.

Cable has said it would accept some form of regulation of subscriber fees as long as such regulation was federal, not municipal. Because consumers are beginning to regard cable more as a necessity, cable sees rate regulation as inevitable. The trick is to find a happy medium: regulation that is sensitive to the needs of those who pay for cable service, and those who provide it.

PUBLIC CONTROL

Public control is a form of social control. Culture, upbringing, education, economic class, religion, gender, and tastes all interact subconsciously to determine people's likes and dislikes. Some citizens express their opinions about programs and commercials through letters to stations, demonstrations, boycotts, or citizen groups.

In a sense, people vote with their wallets. Several groups, notably the Coalition for Better Television, have achieved programming changes by threatening to boycott products advertised in the shows. In the 1980s, for example, Proctor & Gamble withdrew its advertising from fifty episodes of a series because those particular shows contained material seen by the coalition to be "morally objectionable."

If the ratings drop because the public perceives a program to be objectionable, then station revenues may drop, because smaller audiences mean that the station can expect to charge less for spots. For this reason some advertisers maintain "hit lists" of shows on which controversial content may drive away the type of consumer the advertiser seeks. Should the program be a syndicated one, it may be dropped by the station, thus affecting the revenues of the syndicator.

Some groups of citizens may form to protest one single issue, and initiate a heavy mail-in campaign to advertisers or stations protesting that a program did not handle a matter to their liking, such as those concerned with gun control or abortion.

Ethnic groups, sensitive about stereotypical images of them in the media, have banded together and, through various techniques, persuaded the networks to become sensitive to outdated, negative stereotypes—even to the point of challenging station licenses at renewal time, usually withdrawing the challenge if the licensee agrees to their desire for more or better minority programming or increased minority employment at that station.

There are more than 400 *citizen groups*, national and local, which show their concern about the perceived impact of broadcasting on their lives. Major national citizen groups include the National Citizens Committee for Broadcasting, chaired by Ralph Nader, the Office of Communication of the United Church of Christ, and Action for Children's Television (ACT). Founded by a group of Boston parents in 1968, ACT is still concentrating on improving the quality of children's programs, especially today with continued debates about sex and violence on television, as well as about the types of advertising appearing on shows designed for children. Accuracy in Media (AIM), funded mostly by conservative organizations, has successfully used the fairness doctrine and public pressure to force the networks to present more balanced documentaries.

To avoid direct confrontation with pressure groups, today's managers seek input from people in their coverage area and generally try to anticipate and avoid problems when possible. The main difficulty is to determine whether a local group truly represents the community or whether it is a small but vocal splinter.

PROFESSIONAL AND INDUSTRY CODES

National Association of Broadcasters

For several decades, stations which belonged to the National Association of Broadcasters (NAB) followed the NAB's voluntary codes (32 pages long), which contained suggestions about what may be considered favorable or unfavorable in both programming and advertising. Although they never were regulations, they were used as standards, especially in matters of what is good taste, children's programs, and the amount of time that a station could devote to advertising during different times of day. However, the programming aspects of both the NAB television and radio codes were suspended in 1976 when a U.S. federal judge in Los Angeles ruled that the "family viewing" provision of the television code violated the First Amendment, and that the question of prescreening programs for profanity, obscenity, smut, vulgarity, and proper costuming are matters of subjective judgment.

With lack of national program standards for self regulation the networks' standards departments, which evaluate commercials and programs as they are being developed, now exercise general and specific controls over concepts and words to meet the mores of the mass audience. Many stations continue to use the NAB code's standards as the basis of their own programming and advertising policies, while other stations have created their own written policies and guidelines.

The advertising portion of the codes was questioned in 1982 through an antitrust suit by the Department of Justice, which claimed that the time standards of the NAB codes "artificially curtailed advertising, thus depriving advertisers of open competition." As a result, through a consent decree to abolish commercial time standards in the television code, the NAB agreed to put the code aside. However, in 1990 FCC chairman Alfred Sikes called on broadcasters to reestablish the voluntary industry programming aspects of the code, and the NAB was examining the idea. Texts of the suspended codes may be found in the *Broadcasting/Cable Yearbook*, 1982.

Cable's standards are more liberal because their programs are not using the "public airwaves," and people have private choice about what they wish to view. Cable's freedom is pressuring the networks to be less conservative. The matter remains unresolved at present.

Radio Television News Directors Association (RTNDA)

The Radio Television News Directors Association (RTNDA) supports a "Code of Broadcast News Ethics," establishing the right to refuse to distort the news to satisfy the personal views of owners and managers.

SOURCE CONTROLS

Source control may be overt or covert. Public relations activities, stunts, press releases, editorials, pseudo-events such as ribbon cutting at an opening, all provide ways for the source to set the agenda for what is newsworthy. A source is in control when the video, audio, or written story is used on the air exactly as received by the station, because it can be expected that the source's story covers the positive view of the source's opinion. It is source control when a person is given airtime to discuss an opinion if the station does not also seek to air differing viewpoints. Public relations stunts may be used because they are eye-catching but at the same time may be promoting a business, an idea, or a person. Desks of busy reporters are inundated regularly with floods of news releases and publicity handouts. Research shows that because of time constraints, reporters, especially at smaller stations, may use material just as it comes in. Naturally, seasoned reporters and departments check out a source and other angles, rewriting the story to reflect additional information.

Covert control may be more subtle by preventing news from being made public, such as through monopolies, corporate or government secrecy. Some critics even view a president's "fireside chat" a form of control since the speaker has full control over what is said, without a point of view being explored by an interviewer. An advertiser can also be a controlling force in situations when a station may refuse to air a story or program that would displease the advertiser.

SUMMARY

We find a variety of controls discussed in this chapter. The government's role is primarily to assure satisfactory technical standards of equipment used to serve the public, to protect First Amendment rights, and to assure ethical and reasonable business practices. The primary goal of government overseeing is for broadcasting to be open to diverse opinions that reflect the concerns of a democratic society, as a way to prevent the formation of monopolies which may limit opportunities for such diversification.

Guidelines are found in codes formulated by the industry as a form of self-regulation and self-control.

The public may control in several ways: commercial controls resulting in boycotts of selected advertisers; individual expressions of opinion; and formation of pressure groups.

Finally there is the possibility of source control if media do not maintain responsible reporting standards or alternate viewpoints.

Notes

1. *Broadcasting*: 15 January 1990, p. 55.

7

Management Principles and Theories

EFFECTIVE LEADERSHIP

Management is getting things done through other people, and preparing those moving up with managerial skills. Effective leadership balances the needs of management and the needs of employees. Good management sets standards of performance, which should focus on the job, not the individual employee; and both management and employee should be in clear agreement as to what should be expected of both. Matching the right persons with the right jobs is one of management's recurring challenges. Because people come from different lifestyles and organizational experiences, one specific managerial style might not be suited to every individual; consequently, a good manager knows how to be flexible in personnel matters when the situation requires it. Just as every person is unique, every organization also has its own *corporate culture*. As illustrated in Figure 7, corporate culture is that unique culture which results from the blending and interaction of group values, managerial assumptions, organizational practices, financial concerns, and employees' unique personalities, intermixed with the perceptions which each individual brings from home, education, birth order, life experiences, and so forth.

Understanding Organizational Goals

Some people tend to prefer working independently in situations where they can handle several aspects of responsibility and have fewer layers of hierarchy to work through. On the other hand, there are people who are good team workers and who enjoy working on specific aspects of parts of a whole. In corporate work one must be prepared to handle group responsibilities, attend decisionmaking meetings, prepare and respond to reports, take part in informed discussions, and be willing to compromise with a sense of group "oneness" rather than a "me against the world" attitude. Individuals who desire management level positions can learn to handle both team and independent work as they integrate their interests and abilities for the good of the corporate culture.

Motivation

Being able to motivate is the test of leadership. Subordinates want to work for a genuine leader, not a boss. Being able to motivate and to meet organizational objectives are management's two top priorities; and the higher an executive goes up the career ladder, the more that person is judged by success in *delegating responsibility*.

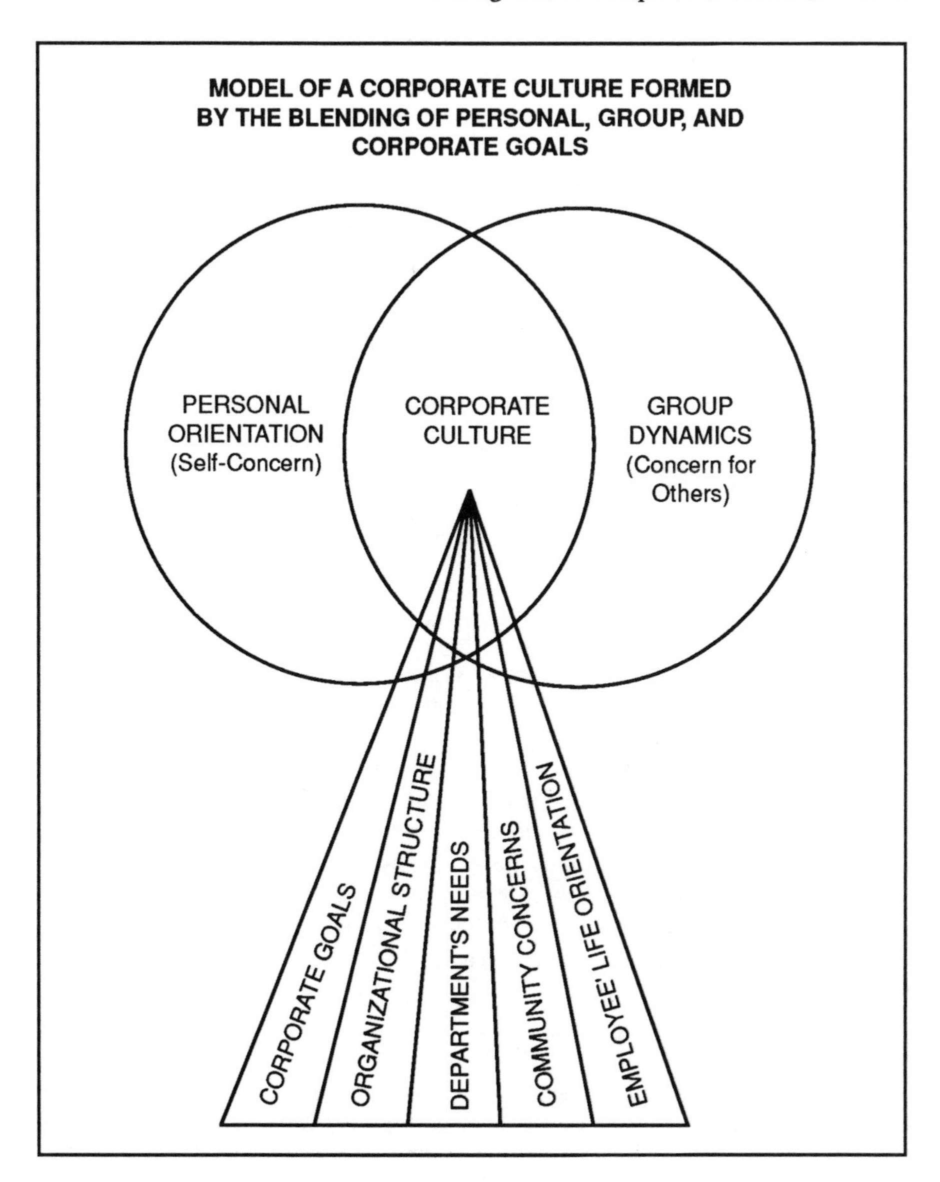

▶ ***Figure 1*** *Cross-cultural influences.*

To achieve motivation and effective organizational results, a manager should first organize a system that enables people to work efficiently. Asking employees their opinions is a measure of management strength, not of weakness. Research shows that the democratic, participative management style leads to the most effective productivity, even though the process takes more time. Once the organization is in place, management should focus on planning future goals after which to set the objectives in words, then try to anticipate obstacles by outlining procedures and, finally, to assign responsibilities. Naturally, as situations change, so should plans.

Problem Solving

A good decision-maker can separate symptoms from causes and, therefore, is less concerned with *who* made the mistake, and more concerned with the *why*. A major rule in business is: find out what's wrong and fix it, find out what's right and reinforce it. But the human factor sometimes complicates this simple rule. There are some co-workers who take everything personally so that when you are trying to find out what happened and why, some people focus immediately on themselves thinking, "Was it something I did? Something I said?"

On the other side, there are people who refuse to accept responsibility for anything whether it is credit or blame. If it was right, they call it luck; if wrong, it was fate. In stress situations the first type finds it difficult to focus on the problem rather than on personalities. And the second type is unwilling to give either positive or negative input. These attitudes are shaped by our early experiences, and we can work to modify them if we recognize them in ourselves.

Achieving Group Consensus

Management that depends upon group participation needs to be able to resolve intergroup disputes. Most organizational personnel problems are caused by open conflict based on a win-or-lose attitude. Robert R. Blake and James S. Mouton identified problem-causing communication types who interfere with achieving group consensus, listed below, in order from bad to good: people withholding to avoid revealing differences; complying, with reservations; glossing over differences, instead of facing them. The next stage toward agreement is compromising by "just going along." The goal of consensus is nearly attained when there is group acquiescence which, hopefully, then leads to full commitment. A committee chair who can recognize these stages and work through them builds intergroup cooperation and a committed corporate culture.

MANAGEMENT THEORIES

Behavioral science concepts applied to management have led to a number of theories about how and whom to manage. Let us examine an overview of the findings.

McGregor's Theories X & Y

Douglas McGregor defines his Theory X as *authoritarian*, which assumes that people do not like to work and must be told exactly what to do. On the opposite end of the scale is Theory Y, which assumes that people have no specific feeling toward liking or disliking work, but develop attitudes toward it while on the job. McGregor finds that under the right circumstances, people seek responsibility instead of avoiding it. Beginners may need and respond best to forms of Theory X, especially if the work is routine or technical. But as they become skilled they are likely to become more effective and responsive under Theory Y management style. According to McGregor, the task of management under Theory Y is to "make the job the stage" to develop in employees competence, self-control, and a feeling of accomplishment.

Sometimes as people move up the ladder they may have to examine and modify their leadership styles. For example, a person who originally headed a department in which direct supervision was necessary may find it difficult to learn to delegate responsibility at the higher levels, and may, as phrased in *The Peter Principle*, have been promoted to his or her "level of incompetence" unless she or he can learn to delegate.

Mastering the art of delegation requires patient, earnest effort and requires conscious selection of the tasks one reserves for oneself, followed by deliberate delegation of everything else to appropriate others in the organization. Executives cannot permit their minds or their desks to be cluttered up.

The delegator needs to be receptive to other people's ideas. Policies must be stated clearly and explicitly. Ideas must be shared. Most importantly, a top executive must establish a means to know what is being done, and when to step in to take action, if necessary.

Expounding McGregor's theories, Rensis Likert's organizational research identifies what he considers to be four basic styles of leadership:

System 1: Exploitive-Authoritative
System 2: Benevolent-Authoritative
System 3: Consultative
System 4: Participative-Group

Likert finds that System 4 is ideal for the profit-oriented and human-concerned organization, but that it takes time and effort to achieve.

The Managerial Grid

Robert R. Blake and Jane S. Mouton developed what they call the managerial grid, which analyzes variations of managerial styles in relation to the extremes of concern for people vs. concern for production. At one extreme is the "country club" manager who coddles employees at the cost of lost production. At the other extreme is the caricature of the old line authoritarian with maximum concern for production and minimum concern for people. Most management styles fall between the two extremes. The grid enables leadership to study management styles in relation to the variables of production and people; and to determine what goals to modify or develop in their own organizations. Workshops help managers discover their current philosophies and management styles.

Management by Objectives (MBO)

In current vogue is Management by Objectives (MBO) conceived by Kurt Lewin and popularized by Peter Drucker. MBO helps a manager to devise long-range goals and strategic plans through a process that establishes goals, decides who will perform the work, and determines methods to oversee work performance and to review the results. MBO is time-consuming and requires commitment and understanding by key managers.

As more knowledge is gained about ways to apply behavioral science findings to management, additional theories and methods will arise to be added to the already vast amount of literature on the subject. Individuals can always update themselves by dipping into the literature on the subject. A number of firms provide seminars for their management teams to help them assess and improve management skills.

CLIMBING UP THE LADDER

An aspiring media manager must be able to function with the creative and procedural factors unique to the profession. A broadcast executive is expected to be knowledgeable about business administration, broadcast law, psychology, sales, programming, technology, and people. Managers must touch people's emotions with their software, yet remain solvent with their hardware.

Broadcast executives are usually people who worked their way up through the ranks within the profession, mainly through sales but also from engineering or programming. Brief internships, workshops, or training alone do not prepare a person for the judgments that must be made at the executive level. Training is essential for entry-level jobs; but education is needed for the person with high managerial goals.

There are organizational goals, department goals, career goals, achievement goals, and personal goals. An individual must be able to determine which goal should be given priority at a particular time; or to figure out how several goals can be achieved through one action. For instance, the "prima donna" type who feels that only he or she has the answer is not likely to move up very far. If, however, that same person can frame suggestions in light of their value to the company's goals, those ideas are much more likely to be considered by upper management. In other words, while it is important to have one's self-esteem, it is also important to refrain from ego trips.

A chief executive officer (CEO) or any other major officer of a corporation that sets out to do the best possible job for the organization's shareholders can succeed only if employees are doing their best and working at the highest rate of efficiency. This does not happen unless workers are well-treated and feel they have a stake in the company. However, statistics show that about one out of every ten stations changes ownership each year. When such changes occur, so also do organizational plans and structures, even in large facilities, which may result in abrupt elimination of certain jobs and even formats and philosophies. The increasing number of corporate mergers and hostile takeovers, while of benefit to shareholders, may lead to decreased concern for the well-being of the staff, as more and more business decisions are made from afar by CEOs who live far outside the area.

There is nothing wrong with worrying about profits; but there is something wrong with worrying *only* about profits.

CHALLENGES TO MEDIA MANAGERS IN THE 1990S

Programming

Jay Kriegel, a CBS executive, thinks that the 1980s were the decade of technology and that the 1990s will be the decade of creativity. With technology outstripping programming, there is a need to create new programming and to do it better. The premium will be on program creators and creative producers to provide product not only for the established stations but also for the multiple cable channels.[1]

In the opinion of Alfred Sikes, FCC chairman, "as distribution channels expand there is room for more people to try new ideas" and that "no longer are people going to be able to depend on ownership or a distribution channel to ensure profitability."[2]

Jack Valenti, president of the Motion Picture Association of America, sees tied in with trends in programming a real fight to keep competition in the marketplace. European countries are banding together to build cable systems and, in his opinion, to keep our programs out.[3]

The battle for viewers in the 1990s will be found on the programming and marketing fronts as cable may not compete with the networks as much as they will be competing with each other.

Technologies

Richard Wiley, former FCC chairman, says that "there will be a broadcasting-cable analog solution to the high-definition television," and thinks there may eventually be a digital fiber-optic solution.[4] Rep. Al Swift wonders if we have already lost the HDTV battle; whether it may be too late for broadcasting to get into it.[5] Because the HDTV question is political and economic as well as aesthetic, it is currently far from unsolved.

The question about permitting telephones to get into cable and into direct-to-home program distribution could provide a means for the networks to go directly to the set, bypassing local stations. If the networks cease to supply programming through local affiliates, the economics of television networks could change, which might affect localism.

Radio

The CBS Radio and Group W heads have reported that 1989 was their best year ever for AM radio, but agree that programming will have to play a major role for an AM revival. However, people must be willing to take a risk and invest a lot of their own time to come up with creative ideas. Anticipated spectrum shifts to allow room for more AM radios will also provide a number of challenges in the 1990s.

Radio is expected to gain from the newspaper industry's current problems with advertising and circulation. RAB president Warren Potash says that perhaps radio has succeeded where newspapers have failed, in that radio is more flexible, cost-efficient, and more localized to serve the advertiser.[6] Alfred Sikes agrees, as he sees the 1990s bringing more competition for broadcasters and more quality material.[7]

Localism

The matter of localism concerns a number of media leaders as well as people in Congress. Al Swift (D) of the House Telecommunications & Finance Subcommittee says, "Every broadcaster should be pounding away on localism—to the public and Congress. I think people literally have no idea what they would lose if they lost those local broadcasters."[8]

Marketplace

Media managers face many challenges in the 1990s. The trend by the FCC to permit the marketplace, rather than government regulation, to make decisions creates uncertainty among media managers about how to serve the public while simultaneously pleasing investors. In addition, the unresolved status of cable, the FCC's financial interest and syndication rules (FIN-SYN), as well as competition from new technologies, calls for creative managements and possible new approaches.

SUMMARY

There are numerous theories of management. The thrust of this chapter is to point out that every organization has its own corporate culture that can be altered by a change of leadership style.

These theories are not unique to broadcasting and media, but they are most appropriate for people who wish to eventually be at the top of the ladder.

One of the most difficult leadership skills is delegation. However, people can modify their management style if they can examine themselves as they appear to people under them, and be willing to seek and take constructive criticism by persons they respect. It may be necessary to change one's management style as one moves into a different setting.

Research shows that most people become station managers between the ages of 35 and 45. Those who have their eyes on that goal should develop a personal timetable with this in mind. Broadcasting and electronic media are exciting, high-pressure fields and, therefore, require relatively young people whose personal energy and abilities can keep pace with the rapidly changing field.

Broadcast leaders are constantly updating themselves by attending workshops and professional meetings. Self-education is essential to keep current and to meet the challenges of the 1990s.

Notes

1. Jay Kriegel, Speech RTVMP Department, 16 November 1989.
2. Alfred Sikes, "1990 The New Decade," *Broadcasting*, 1 January 1990, p. 38.
3. "1990 The New Decade," *Broadcasting*, 1 January 1990, p. 39.
4. Ibid.
5. Ibid.
6. Ibid., p. 49.
7. Ibid.
8. *Broadcasting*, 27 November, 1989, p. 48.

Employment in Electronic Media

This chapter explores employment possibilities in electronic media, focusing primarily on managerial positions but also providing an overview of responsibilities and operations at lower level positions. Because there are many opportunities for horizontal employment by moving from one medium to another, the management principles and practices apply not only to broadcasting and cable but also to related media and professions.

CAREER PREPARATION

Education

Although some broadcasters would rather train entry-level employees themselves, a college degree in areas of telecommunications, journalism, business, marketing, or advertising will enable a person to advance more rapidly beyond positions which require only training. A college degree sharpens one's reading, writing, thinking, and speaking skills and also provides professional background and knowledge.

The minimum educational requirement for entry-level jobs is a high school diploma, but advancement opportunities are usually more limited with such preparation. The same can be said about some technical positions. Technical training in electronics is essential for entry jobs in broadcast engineering departments. It is recommended that a person interested in a technical career take appropriate courses in electronics at an accredited technical institute and then gain experience working for radio or television stations. Technical education and training in an accredited college is preferred. Lists of schools and their degree programs are found in *Broadcasting/Cable Yearbook.*

Although a number of managers in good markets got their start at small stations and moved up, those who did were mainly in sales, a field which requires knowledge of people as well as business.

Personal Traits

There are positions in electronic media to match a variety of personalities. People with a creative bent are found employed as copywriters, scriptwriters, announcers, music directors, and television directors. Because creative personalities are likely to look at situations from many perspectives, a creative person can apply that skill at management levels to solve problems.

All individuals planning careers in media must have good health, be able to handle stress and to adapt to last-minute pressures and changes. Candidates must have intellectual curiosity, be able to get along with other people, and, most of all, be flexible. As Roberta Flack said, "An employee must remember that at times one must accept unpopular changes or restructuring because it is normal for businesses to make changes they see as having a better economic impact on the operations. No employee may expect to be pleased with everything."[1]

Basic Skills

It is clear that every media professional's life will be touched by the typewriter, word processor, and the computer. The sooner one learns how to operate these basic tools, the better. No longer are these the province only of secretaries. If it is not possible to take courses to learn to upgrade these skills, people can get the most current books available to learn terminology and concepts. Some public libraries offer personal computer centers where people can sign up and practice in their free time.

If you can arrange to take courses, check the business section of the newspaper for advertisements about classes in the city, or check with your local community college to inquire about minicourses on the introduction to personal computers and various software. Some firms will pay for tuition for such classes. Then be certain to add to your résumé the names of software packages you are familiar with.

Computers in the Media

Although it can be argued that automation decreased the number of personnel necessary to operate a station, it has influenced the growth of the number of radio stations. Today nearly all of the 10,600 radio stations are operated more efficiently thanks to automation and, consequently, more people are working in the industry. But since automation does the routine things that a machine can do best, it saves the people for something more important. And while computer technology and automation have taken the sweat and tedium out of many jobs, work has become far more complex and mentally demanding. No longer can workers count on performing the same task day in and day out. Now they must handle a variety of skills and make snap decisions in unpredictable situations. This leads to challenging adjustments for managers as well as employees.

Managers who must delegate more decisionmaking may feel threatened about relinquishing power. Not all workers have analytical skills or are adept at problem solving nor can work as a team. People who have never before been asked for an opinion suddenly have to make decisions, and sometimes it frightens them. If a person has been conditioned to be fairly passive and supervised, it is a real change in the rules to be told to take responsibility. It is top management's job to be aware of these factors and to have in-house ongoing training to help people adapt.

INSIDE THE STATION: JOB RESPONSIBILITIES

Although this book's focus is upon management, it is important for future managers and also would-be employees to have an overview of responsibilities related to particular areas and departments. The following section may help a newcomer assess the various opportunities found in the different departments.

General Staff

Without good support staff there is no smooth operation. Needed to coordinate station activities are receptionists, secretaries, and bookkeepers, as well as department managers. People in such positions can learn the broad workings of the industry and the particular station, and can find it excellent preparation for that first step up the ladder toward managerial responsibilities.

Staff Announcer

The staff announcer typically reads commercial copy, introduces live and recorded programs, gives station identification, and makes promotional and public service announcements. In small stations this is usually handled in conjunction with console operation or disc jockey functions or both. With automation the announcer may prerecord all announcements during assigned production times.

Production Director

The production director plans and supervises the production of a show, or a series of shows, and also, at times, commercial or promotional announcements. In television this job is more demanding as it involves coordination of the various elements of a program, which includes selection of film, scripts, music, and budget.

Production Manager

The production manager reports directly to the program director and handles many of the details involved in producing all the station's live programs. This person selects personnel, space, and determines equipment requirements, working closely with the production director.

News Director

Responsible for gathering and disseminating news, the news director determines the news policy of the station, supervises news personnel, and may also serve as a working reporter.

Newscaster

The newscaster functions as a reporter, collecting local news and rewriting selected stories from the wire services. The newscaster must have good air skills, and be able to write copy that sounds good on the air.

Promotion Manager

In radio the promotion may be handled in the traffic or program department. In television most large stations have promotion departments headed by a promotion manager. This department's responsibility is to secure good publicity for the station, its programs, and its talent. The department also works with the planning and layout of advertising campaigns and promotional activities aimed to capture the station's audience. In addition, the promotion department may handle sales promotions, plan the layout of advertising for trade journals, and produce the sales brochures and other materials used by the sales department.

Traffic

The traffic department handles daily scheduling of facilities, personnel, program material, and announcements, which are scheduled on the daily program log prepared by the traffic department. Except in very small stations, the log is usually done on computer. Traffic also monitors whether commercials were run as scheduled; and arranges "make-goods" if spots failed to be aired. Traffic also keeps sales department personnel informed about availabilities or commercial openings in the schedule. Traffic usually fills empty slots with public service or promotional announcements.

Continuity writers usually work in the traffic department and are responsible for writing local commercials, public service spots, or promotional material. Continuity may also write portions of local programs.

Sales

Broadcasting sales is challenging in that it is the art of selling intangibles. A successful salesperson must have extensive knowledge and understanding of the broadcast product, audience analysis, and a potential client's needs—and the ability to put it all together creatively. People in sales can find such employment extremely lucrative if they find they have a talent for selling. Understandably, a person in sales needs to like people and to have physical and mental stamina to keep up with the position.

Business

Broadcasting is a business. Every industry needs people who understand what it takes to make a business run. The same applies to broadcasting, which requires managers, financial experts, and departmental administrators—skilled executives with sharp minds who can lead a group of people to successful goals.

Technical

Electronics is one of the most rapidly changing fields in the history of mankind. The need for competent, technically-oriented people is increasing as the field turns more than ever to automation and use of computers. Not only are technicians needed to maintain and repair equipment, but also to be innovative in adapting technologies to the particular station's needs. Thus a good technician and engineer is more than a nuts-and-bolts individual; creativity and eventually supervisory skills are also required for upward mobility.

Other/Cable

There are a number of other staff positions in each of the above-mentioned areas. The preceding were detailed to provide an overview of personnel in a typical station. Cable systems have similar organization structures, only with different titles, namely: system manager, office manager, marketing director, chief technician, local-origination coordinator, and installer.

ENTERING THE MARKET

Finding a Job

Finding a job in broadcasting, especially television, is not easy. Because television is both glamorous and a highly paid industry, it attracts many more people than there are jobs. Competition is tough. And finding that first job is not easy, even for people with college degrees. So how does one break in? Those who get their foot in the door are those willing to take menial jobs for low pay and work themselves up. If possible, getting summer internships at local radio, television, cable, and production systems is a way to learn about the real world and to make valuable contacts. Occasionally, you may be the lucky one to get a job offer from the station or firm you interned with. Sometimes the internship may be in an allied field, such as a local newspaper. Take it! Many people cross over into other media, so don't turn down an opportunity, even if it is not your first choice.

If you cannot get an internship, offer your time free on weekends. You are making an entry of a sort. And you are also discovering what you like or dislike. While in school volunteer to help organizations with their publicity; videotape activities for the yearbook; work at the student radio or television station. Look for opportunities to show your skills. If you are creative you might even devise a program you can write and appear in on the local UHF or cable station, bringing in your friends as appropriate talent. This would demonstrate your leadership potential as well as your talent, and will look good on your résumé. Be willing to work with little theater groups on props and lights if production interests you. At the same time, you will be meeting people with similar interests and again, making useful future contacts.

When the time comes for the search for the real job, first analyze all the contacts you have already made and ask them if they have heard of openings. Let them know of your availability. Do not expect to be paid much at first. It is a blow to a college student to find out the job may pay no more than a fast-food service position, but the point is to get inside.

In addition to scrutinizing advertisements in local papers, look at the ads in the trade magazines and newspapers (if you are willing to live anywhere). Make appointments at places where you would like to work. Even if you do not get the first jobs sought, you have gained experience in handling yourself during the interview. When possible, get information in advance about the firm at which you are applying. And flood the appropriate firms and also friends in media with copies of your résumé.

Some of the best job leads come from a friend, or a parent's friend, who knows of an opening and would like to get your résumé off his desk. Do not sit back and wait for a reply to just one letter of inquiry. The rule of thumb is, feel lucky if you get one interview for every 100 inquiries. Finally, if you are a student, start your search at least six months before graduation. Even 12 months is not too soon.

One of this author's students wanted to work for a particular network when she graduated. Beginning in her freshman year she dropped by, with an appointment, to see the network's personnel manager saying something like, "I know this is early but I wanted to get acquainted with you and have you get to know me because I would like to work here when I graduate." And every fall, Christmas and spring break she repeated the visit, talking about what she was studying, commenting on current happenings in the field and, especially, that network. Upon graduation she applied for an official interview and, armed with a résumé, went through the entire process despite the fact that she had been told there was little chance of a beginner working with the network. Her broadcast skills were extensive and she was good at writing and organizing. But she had reluctantly also taken typing and shorthand. Passing the required tests in shorthand and typing, she was placed in the secretarial pool. Just three weeks later there was a vacancy working for the top morning show where she started as a gofer and eventually worked up to be the top research assistant for Barbara Walters, flying in advance to foreign locations to prepare for the upcoming broadcasts, doing research on the personalities and background, and occasionally filling in on the air. She prepared herself in advance, and her intelligent persistence opened the magic door.

Preparing the Résumé If you are unfamiliar with résumé writing get a good book in the library or bookstore. Ask a teacher or professional friend to critique your first efforts. It is not easy to write a résumé. You know all about yourself, but you have to present yourself on just one page of paper. Choose a résumé format you like, then write and rewrite. Use only information that will help show your background and special experience and abilities for each particular job. Mass-produced résumés are too general. Résumés should not be photocopied. Ideally, each should be an original. With the aid of a word processor you can prepare an original each time, changing only the portions which can slant the résumé to each firm's uniqueness.

A job title should head the résumé. Employers do not want to go on a fishing expedition with you to find out what you want to do. A job title tells them. In addition to the obvious name, address, and phone number, list your experience, education in related activities as briefly as you can, with the most recent listed first and each category.

You may have more related activities than you realize. Beyond experience in the media, list chairmanships or campus offices you have held. Include high school club offices as well as work in outside organizations such as Explorer Scouts and any well-known group that requires organizational skills, discipline, and shows you have had experience working with people and groups.

Do not overlook mentioning part-time jobs. They indicate your are willing to work and are dependable. You may want to ask a close friend to help you discover your personal assets; many persons do not recognize all their own strengths. And even after you get a job, keep working on that résumé; updating it continually. You will find it much easier to prepare new résumés and will be ready in case the ax falls because of something beyond your control, such as a corporate takeover or new management plan. Always plan ahead!

The Audition Tape If you are auditioning as talent, or wish to illustrate your production skills, you may wish to submit a 5-minute audio- or videotape. Keep it brief. A skilled employer can judge in the first 2 minutes if you have talent appropriate for that particular firm. Try to demonstrate your versatility and accompany the tape with a perfectly typed summary listing what the tape contains, the length of each segment, and if you wish to, a preface stating why you included it. Be sure that it demonstrates your knowledge of appropriate format as well. Make certain your name and address is on the tape, and that it is a standard speed and format. Generally the tape will not be returned; but if you feel it may help you get the job, it is a worthwhile investment.

The Interview The interview actually begins long before you get to the reception room. Have at least two copies of your résumé handy, one for you to refer to and one to leave with the interviewer. Be prepared with a good pen to fill in the application forms carefully. Carelessness can show your failure to pay attention to detail and result in your application's being piled on the reject list. Also learn something in advance about the company structure, philosophy, and holdings. If there has been no advertised opening, you may wish to plan how your skills and the firm's goals mesh.

Appearance is most important for that first impression. One should dress appropriately for the job and be well-groomed. During the session act naturally, and do not be offended by a question that seems irrelevant to you. The interviewer may be seeking some quality necessary for the job, or may be testing you and your temper.

Listen carefully to each question. Do not answer before you contemplate what is asked. Remember, it is less important to answer at once than it is to answer fully. It is perfectly acceptable to think a moment to pull your answer together; do not let anxiety rush your answer. Sometimes an interviewer allows as much as a minute of silence, partly to test the subject, and partly to provide time for additional remarks or questions.

If you have questions you wish answered, be sensitive to when and how to ask them.

During the interview don't be boastful, but know and point out your strengths. As a note of caution keep in mind that some station executives are reluctant to hire those college graduates who enter with a superior attitude, and who seem more concerned about showing off than in listening and thoughtfully answering questions that are asked. Even if you are never asked the formerly familiar question "Where do you want to be in 5 years?" you can mentally prepare an answer as a way to think of what you actually want.

Corporate interview techniques are changing, spurred by increased costs of employee turnover, training, and termination. As more employees switch careers, employers are seeking ways to ascertain how a checkered past can fit a particular future. In addition, interviews are only part of the hiring process in many firms today, being supplemented by personality tests, task simulations, and so forth. The questioning might focus on how the interviewee handled a particular situation. And interviewers are trained to ask follow-up questions. If candidates list specific extracurricular activities, they might be asked what was really done and how. If in charge of a dance committee, for example, the questions might be, How many people worked under you? How did you organize the group? What worked well and why? What did not work and why? Sometimes recognizing why things do *not* work indicates analytical ability in self-assessment, plus problem-solving skills.

Interview Follow-Up Repeat interviews are common in big firms, so don't be offended if you are told you'll hear later and then are asked to come back. It is appropriate for you to ask when your application will be acted upon. And you may ask if you may call in a few days to find out if a decision has been made and then phone on the day agreed upon. Perhaps you were the second choice, but the first choice turned down the job offer. In that circumstance you might get the job by the initiative of a follow-up phone call. Always write a thank you letter immediately after the interview and mention that you hope to hear about the decision soon. Be persistent in a subtle way. And always focus on reasons why you can fill the *specific* job for which you are applying. Make a distinction between what you *have to offer* and what you *want.*

Minority Employment Opportunities

As a partial response to the concerns expressed in the 1968 *Report of the National Advisory Committee on Civil Disorders*, usually referred to as "The Kerner Report," the FCC voiced policies and principles to prevent employment discrimination against minorities. As a result, the commission adopted rules which not only forbade discrimination on the basis of race, color, religion, or national origin, it also required "equal employment" by all licensees or permittees to all qualified persons. A decade later the Federal Communications Minority Ownership Task Force reported that despite the fact that minorities constituted approximately 20% of the population, they controlled fewer than one percent of the 8500 commercial radio and television stations operating in the United States at that time.

In the 1970s not only ethnic minorities but also women were seeking to break through the "glass ceiling" and move up to more responsible positions. Sensitive to this, the FCC then included women in its EEO and Affirmative Action guidelines. That means a station must go beyond nondiscrimination and actually develop a plan to seek out and retain qualified women and minority applicants.

While training programs are not mandatory, the model EEO policy suggests that such programs would help a station meet EEO requirements. Many large stations have developed training programs. Others provide scholarships and internships for minorities enrolled in local colleges and universities.

In general the ideal goal is for the station's personnel to reflect the percentages of minorities and women in their markets. Stations must compile and report annual figures on the number and percentages of women and minorities employed and must also describe any promotional policies or practices which have benefitted these two groups.

The Cable Communications Act of 1984 includes similar guidelines for cable systems and sets parity percentages of employment in ratio to the number employed in different categories.

To prevent discrimination during the employment process, for example, it is illegal for an interviewer to ask questions soliciting information about race, creed, age, ethnic origin, marital status, number of children, or sexual orientation. The important criterion is that unless a question relates directly to the job or to specific needs, it should not be asked and, if asked, need not be answered.

While the black movement gave the first impetus for entry of ethnic minorities into broadcasting and broadcast management, it set the pattern for women to follow a decade later. In 1990, white males still filled the majority of key management positions in commercial broadcasting and in media corporations. Women have been moving up, however, especially in public television decisionmaking jobs. The highest percentage of women in management is found in the cable industry as System Managers and Marketing Directors.

The employment of members of minority groups has increased in both commercial and public television in the last 10 years, and today they hold an average of 18% of all jobs in commercial television, and approximately 20% in public television. Women and minority personnel hold a large percentage of entry-level clerical and secretarial jobs, but are slowly moving up into higher level positions. An excellent reference book which lists career requirements, salary range, prerequisites, and advancement prospects is *Career Opportunities in Television and Video*, cited in the bibliography.

Unions

The larger the station and the market, the more likely it will be that employees will have to join certain unions related to their specialization. Broadly, there are two categories of groups: (1) creative/performing and (2) crafts/technical.

Most of the creative/performing unions now in broadcasting originally came from the state and motion pictures, such as the Writers Guild of America (WGA), American Guild of Variety Artists (AGVA), and the Screen Actors Guild (SAG). The first broadcasting union to arise was the American Federation of Television and Radio Artists (AFTRA). Many actors belong to both SAG and AFTRA.

Technical unions in broadcasting were begun in the 1920s and wield very strong power, even today. The first union purely for broadcasting technicians was the national Association of Broadcast Engineers and Technicians (NABET), which later changed the word *engineers* to *employees.* Coming to the field of television out of motion pictures is the International Alliance of Theatrical Stage Employees and Moving Picture Machine Operators (IATSE). Electricians belong to the International Brotherhood of Electrical Workers (IBEW).

Unions have gained increased wages for their members. They also have, through their strikes, adversely affected writing and production at major studios and networks, often creating delay of plans for entire seasons.

Salaries

As a rule the larger the market, the higher the salary, with people in the top ten markets earning the most. But not everybody can aspire to match Dan Rather's $3 million a year, Bryant Gumbel's $2 million, or the $1 million plus earned by Connie Chung, Barbara Walters, Jane Pauley and Deborah Norville. Those are the exception, and the lucky ones.

Television salaries are normally higher than comparable positions in radio. At the management level the national median earnings of a general sales manager in 1988 was $81,000 in television and $48,000 in radio. The television national sales manager's median earning was $66,000 in television and $58,000 in radio. Salaries for television department heads ranged from $50,000 to $28,000, while in radio the top average was $29,000 for television compared to $18,000 in radio. The average for a general manager was $103,000 in television. Keep in mind that an average is a midpoint figure, and that there are higher and lower salaries averaged in to find the medium.[2] Below management levels the difference between wages and workload in a nonunion shop *vs.* a union shop are extensive, especially in engineering. People who remain in radio generally do it because they enjoy it, despite the lower financial compensations.

News anchors, who usually earn even more than news directors, earned a median of $55,000 in television and $20,000 in radio in 1988.[3] An annual "visibility" study of television by Joseph Foote, as reported in *Media Report to Women*, reveals that of the 50 correspondents seen most often on the newscasts aired by the three major networks in 1988, only six were women. However, women increased their share of news directorships in 1988, mainly at network affiliates, where an estimated 82 women were news directors, according to Vernon Stone, director of research services for the RTNDA (as reported in the August 1989 issue of RTNDA's *Communicator* magazine). Stone's study revealed that about 50 women held such positions at independent television stations. In commercial radio, according to Stone, news directors included one woman for every three men.

Of the nation's 740 news directors 115 were white women and 58 were minorities.

Advancement Prospects

One way to assess advancement prospects is to realize there are more opportunities in the industry with the most units. Thus, with only 1395 operating television stations on the air, there are only that many opportunities for one particular job title at each station. Radio's some 10,461 stations offer ten more opportunities for every one of the same position in television. Cable systems reached nearly the 10,000 mark in 1990, and, therefore, offer that many more electronic media type opportunities.

Persons interested in detailed job descriptions and current openings in broadcasting should start with a study of the classified ads appearing weekly in *Broadcasting* magazine. A close examination of those pages provides a comprehensive portrait of job requirements and salaries. In addition to opportunities in cable, Multipoint Distribution Service (MDS), Subscription (over-the-air) pay television (STV) and the developing satellite systems mentioned in Chapter 2, there are career opportunities in other consumer electronic industries and video marketing.

Positions in any of the professional associations can lead to managerial experiences. These may include organizations such as the National Association of Broadcasters, National Association of TV Program Executives, Cabletelevision Advertising Bureau, Television Bureau of Advertising, Radio Advertising Bureau, Association of Independent TV Stations, Electronic Media Rating Council, Radio-TV News Directors Association, National Cable Television Association, State Broadcasters Associations, union and labor organizations, as well as station and CATV brokerage firms.

There are also a number of government organizations such as the U.S. Information Agency and the Voice of America, which employs large numbers of people in many parts of the world. Private industry with corporate video departments offers a variety of production type positions, as do hospitals and other health organizations. Persons interested in the field of education might find satisfactory work as director of media services.

Public relations offers a wide choice of media-related jobs as does the field of advertising. Working with large advertising agencies enables one to use or to develop many media-related skills as illustrated earlier in this book.

Not to be overlooked are positions in marketing, points out Barry Salbert, management consultant, who says, "Radio is in dire need of legitimate marketing and sophisticated strategic planning" and needs more than research or knocking on doors for an order. Salberg stresses that "advanced marketing methodology and formal strategic planning goes beyond just sales and programming, or even promotion; yet it encompasses the best of all three."[4]

Job Termination

Turnover in media jobs is higher than in most fields. Radio stations generally have a higher turnover in nonmanagement positions than does television. But in television, there is a higher rate of managerial turnover than in radio.

While some employees are terminated for poor work or work habits, most terminations, about 36% in radio and television, result from managerial policy changes such as cutting back staff size, change of ownership or management, format changes in radio, and station automation. Because more television stations are unionized than are radio, job termination in television is more difficult to handle from a managerial viewpoint.

Legal grounds for job termination include such actions as possession of a dangerous or lethal weapon, deliberately tampering with company records, abuse of alcohol when it affects the job, excessive unexplained absences, misuse of company property, insubordination, stealing, and use or sale of illegal drugs. Additional serious infractions that can result in dismissal are failure to observe safety rules, committing a crime which results in a felony conviction, or willful and knowing disregard of FCC regulations.

Large broadcasting and cable firms, MSOs, group owners, and conglomerates publish employee handbooks which explain the policies and practices of each organization and cover all aspects of organizational structure: departments; employee relations concerning promotion, demotion, and dismissal; and grievance procedures. The handbooks also spell out details of fringe benefits, pensions, vacations, educational opportunities, what the employee should know about the FCC, plus how the station can help the employee. Such firms also often publish "house organs" in the form of newspapers or newsletters, with up-to-date information about the firm's successes and plans as well as news about the employees and any changes that may affect them.

Obviously the picture changes the higher one goes up the job ladder, and as one moves into management, that person also must take on the mirror responsibility of informing and training new employees.

SUMMARY

Broadcasting offers challenging opportunities to the person with intellectual curiosity, a willingness to learn, good health, and flexibility. There are positions to match a variety of personalities and talents. Because of intense competition for entry-level jobs, the person with a college education has a better chance of being hired, and the degree need not be strictly in broadcasting.

An individual with aspirations for future management positions should, however, be willing to accept any first job assignment, which may often be in clerical or gofer activities. Essential skills for media employment include ability to write well, to be able to type, and to work with computers.

This chapter provided a summary of responsibilities held by employees in the various departments in the station to illustrate the numbers and types of activities essential to a station's operation.

For the benefit of the applicant seeking the first job, the section on how to enter the market covered basic job-hunting skills in addition to specific matters related only to broadcasting.

Qualified minorities and women may receive special consideration for employment, particularly if a station does not already have enough people to reflect the percentage of minorities and women in its market. Many stations have instituted special training programs or scholarships for minorities who show aptitude and talent for the field.

A person moving into larger stations and markets may be required to join a union, especially if that individual works in technical or creative/performing fields.

Television salaries are generally higher than comparable positions in radio, with people in the top ten markets earning the most. The range of salaries paid to people in management positions offers a broad picture of what a person may earn, but each station is different. The figures cited suggest that the earnings are competitive with those being paid to people holding managerial responsibilities in other fields.

The final section of this chapter listed the numerous job opportunities which exist in organizations outside of broadcasting stations and includes professional associations, government organizations, private industry, public relations, advertising, and marketing.

Notes

1. United States Government, National Advisory Commission on Civil Disorders, *Report of the National Advisory Commission on Civil Disorders*, (New York: Dutton, 1968).
2. *Working Woman*, January, 1989, p. 73.
3. Ibid.
4. *Broadcasting*: 29 January 1990, p. 17.

BIBLIOGRAPHY

Abel, John, et al. *Radioutlook: Forces Shaping the Radio Industry*. Washington, D.C.: National Association of Broadcasters, 1988.

Bittner, John R. *Broadcasting and Telecommunication: An Introduction*. 2d ed. Englewood Cliffs, NJ: Prentice-Hall, Inc., 1985.

Blake, Robert, and Jane Mouton. *The Managerial Grid*. Houston, TX: Gulf Publishing, 1964.

Broadcasting Bibliography: A Guide to the Literature of Radio & Television. 2d ed. Washington, D.C.: National Association of Broadcasters, 1984.

Broadcasting/Cable Yearbook. Washington, D.C.: Broadcasting Publications, Inc., annual.

Brooks, Tim, and Earle Marsh. *The Complete Directory to Prime Time Network TV Shows 1946–Present*. 4th ed. New York: Ballantine, 1988.

Cassata, Mary, and Thomas Skill. *Television: A Guide to the Literature*. Phoenix, AZ: Oryx Press, 1985.

Cable TV Fact Book: Guide to Cable America. New York: Cable Television Advertising Bureau, Inc., 1984.

Code of Federal Regulations. Title 29: Labor. Washington, D.C.: U.S. Government Printing Office, annual.

Code of Federal Regulations. Title 47: Telecommunication. Washington, D.C.: U.S. Government Printing Office, annual.

Corporation for Public Broadcasting. Annual Report. Washington, D.C.: CPB, annual.

Eastman, Susan Tyler, Sydney W. Head, and Lewis Klein. *Broadcast Programming: Strategies for Winning Television and Ratio Audiences*. Belmont, CA: Wadsworth Publishing Co., 1981.

______ and Robert A. Klein, eds. *Strategies in Broadcast and Cable Promotion*. Belmont, CA: Wadsworth Publishing Co., 1982.

Employee Compensation and Fringe Benefits Report: Radio. Washington, D.C.: National Association of Broadcasters, annual.

Employee Compensation and Fringe Benefits Report: Television. Washington, D.C.: National Association of Broadcasters, annual.

Equal Employment Opportunity Trend Report. Washington, D.C.: Federal Communications Commission, annual.

Head, Sydney W., and Christopher H. Sterling. *Broadcasting in America; A Survey of Electronic Media*. 5th ed. Boston: Houghton Mifflin Co., 1987.

Heighton, Elizabeth J. and Don R. Cunningham. *Advertising in the Broadcast and Cable Media*. Belmont, CA: Wadsworth Publishing Co., 1984.

Hills, Jill. *Deregulation Telecommunications; Competition and Control in the United States, Japan and Britain.* Westport, CT: Greenwood/Quorum, 1986.

Howard, Herbert H. and Sidney L. Carroll. *SMATV: Strategic Opportunities in Private Cable.* Washington, D.C.: National Association of Broadcasters, 1982.

______, and Michael S. Kievman. *Radio and TV Programming.* Columbus, OH: Grid Publishing, Inc., 1983.

Hudspeth, Delayne R. and Ronald G. Brey. *Instructional Telecommunications: Principles and Applications.* New York: Praeger, 1986.

Lavine, John M. and Daniel B. Wackman. *Managing Media Organizations: Effective Leadership of the Media.* New York: Longman, 1988.

Likert, Rensis. *New Patterns of Management.* New York: McGraw-Hill Book Co., 1961.

______. *The Human Organization.* New York: McGraw-Hill Book Co., 1967.

MacGregor, Douglas. *The Human Side of Enterprise.* New York: McGraw-Hill Book Co., 1982.

McCavitt, William E. and Peter K. Pringle. *Electronic Media Management.* Boston: Focal Press, 1986.

Nielsen Report on Television. Northbrook, IL: A.C. Nielsen Co., annual.

O'Donnell, Lewis B., et al. *Radio Station Operations: Management and Employee Perspectives.* Belmont, CA: Wadsworth Publishing Co., 1989.

Policies for Regulation of Direct Broadcast Satellites. Washington, D.C.: Federal Communications Commission, 1980.

Poltrack, David. *Television Marketing: Network, Local, and Cable.* New York: McGraw-Hill Book Co., 1983.

Radio Financial Report. Washington, D.C.: National Association of Broadcasters, annual.

Radio Program Department Handbook: A Basic Guide for the Program Director of a Smaller Operation. Washington, D.C.: National Association of Broadcasters, annual.

Radio Today. New York: Arbitron Ratings Co., annual.

Reed, Maxine K., and Robert M. Reed. *Career Opportunities in Television, Cable and Radio.* New York: Facts on Files, Inc., 1986.

Report on Syndicated Programs. Northbrook, IL.: A.C. Nielsen Co., annual.

Rogers, Everett M. *Communication Technology: The New Media in Society.* New York: Free Press, 1986.

Routt, Ed., James B. McGrath, and Fredric A. Weiss. *The Radio Format Conundrum.* New York: Hastings House Publishers, Inc., 1978.

Sherman, Barry, L. *Telecommunications Management: The Broadcast and Cable Industries.* New York: McGraw-Hill Book Co., 1987.

Smith, F. Leslie. *Perspectives on Radio and Television: Telecommunication in the United States.* 2d ed. Harper and Row, Publishers, Inc., 1985.

Station Policies and Procedures: A Guide for Radio. Washington, D.C.: National Association of Broadcasters, 1983.

Sterling, Christopher H., and John M. Kittross. *Stay Tuned: A Concise History of American Broadcasting.* 2d ed. Belmont, CA: Wadsworth Publishing Co., 1989.

Syndicated Program Analysis. New York: Arbitron Ratings Co., annual.

Television Financial Report. Washington, D.C.: National Association of Broadcasters, annual.

Vogel, Harold L. *Entertainment Industry Economics: A Guide for Financial Analysis*. New York: Cambridge University Press, 1986.

Warner, Charles. *Broadcast and Cable Selling*. Belmont, CA: Wadsworth Publishing Co., 1986.

Webb, G. Kent. *The Economics of Cable Television*. Lexington, Mass.: Lexington Books, 1983.

Periodicals

Advertising Age. Chicago: Crain Communications, Inc., weekly.

BM/E: Broadcast Management/Engineering. New York: Broadband Information Services, Inc., monthly.

Broadcasting. Washington, D.C.: Broadcasting Publications, Inc., weekly.

CableVision. Denver, CO: Titsch Communications, Inc., weekly.

Electronic Media. Chicago: Crain Communications, Inc., weekly.

Media Line. Washington, D.C.: National Black Media Coalition, bimonthly.

Media Report to Women. Silver Spring, MD: bimonthly.

The Professional Communicator. Austin, TX: bimonthly.